T0329715

Advanced Battery Management Technologies for Electric Vehicles

Automotive Series

Series Editor: Thomas Kurfess

Advanced Battery Management Technologies for Electric Vehicles

Rui Xiong
Beijing Institute of Technology
China

Weixiang Shen
Swinburne University of Technology
Australia

The right of Rui Xiong and Weixiang Shen to be identified as the authors of this work has been asserted in accordance with law.

Registered Offices
John Wiley & Sons, Inc., 111 River Street, Hoboken, NJ 07030, USA
John Wiley & Sons Ltd, The Atrium, Southern Gate, Chichester, West Sussex, PO19 8SQ, UK

Editorial Office
The Atrium, Southern Gate, Chichester, West Sussex, PO19 8SQ, UK

For details of our global editorial offices, customer services, and more information about Wiley products visit us at www.wiley.com.

Wiley also publishes its books in a variety of electronic formats and by print-on-demand. Some content that appears in standard print versions of this book may not be available in other formats.

Library of Congress Cataloging-in-Publication Data

Names: Xiong, Rui, author. | Shen, Weixiang, author.
Title: Advanced battery management technologies for electric vehicles / Rui
    Xiong, Beijing Institute of Technology, China, Weixiang Shen, Swinburne
    University of Technology, Australia.
Description: Hoboken, NJ : John Wiley & Sons, Inc., [2019] | Series:
    Automotive Series | Includes bibliographical references and index. |
    Identifiers: LCCN 2018044881 (print) | LCCN 2018046237 (ebook) | ISBN
    9781119481676 (Adobe PDF) | ISBN 9781119481683 (ePub) | ISBN 9781119481645
    (hardcover)
Subjects: LCSH: Electric vehicles–Batteries.
Classification: LCC TL220 (ebook) | LCC TL220 .X56 2018 (print) | DDC
    629.25/024–dc23
LC record available at https://lccn.loc.gov/2018044881

Cover Design: Wiley
Cover Images: © solarseven/Shutterstock, © 3DMI/Shutterstock, © Ungor/Shutterstock, © buffaloboy/Shutterstock

Set in 10/12pt WarnockPro by SPi Global, Chennai, India

Printed in the UK

Contents

Biographies

Rui Xiong received his MSc degree in vehicle engineering and PhD degree in mechanical engineering from Beijing Institute of Technology, Beijing, China, in 2010 and 2014, respectively. He conducted scientific research as a joint PhD student in the DOE GATE Centre for Electric Drive Transportation at the University of Michigan, Dearborn, MI, USA, between 2012 and 2014.

Since 2014, he has been an Associate Professor in the Department of Vehicle Engineering, School of Mechanical Engineering, Beijing Institute of Technology, Beijing, China. Since 2017, he has been an Adjunct Professor at the Faculty of Science, Engineering and Technology, Swinburne University of Technology, Melbourne, Australia. He has conducted extensive research on electric vehicles and hybrid electric vehicles, energy storage and battery management systems, and authored more than 100 peer-reviewed articles and held ten patents in the relevant research fields.

Dr Xiong was a recipient of the Excellent Doctoral Dissertation from Beijing Institute of Technology in 2014, and the first prize of the Chinese Automobile Industry Science and Technology Invention Award in 2018. He received the 2018 Best Vehicular Electronics Paper Award recognizing his paper as the best paper in Vehicular Electronics that had been published in the *IEEE Transactions on Vehicular Technology* over the past 5 years, and the Best Paper Awards from *Energies*. He is as an Associate Editor of *IEEE Access*, and an Associate Editor of *SAE International Journal of Alternative Powertrains*. He is also serving on the Editorial Board of *Applied Energy, Energies, Sustainability*, and *Batteries*. He was a conference chair of the *International Symposium on Electric Vehicles (ISEV2017)* held in Stockholm, Sweden, 2017, and a conference chair of the *International Conference on Electric and Intelligent Vehicles (ICEIV 2018)* held in Melbourne, Australia, 2018.

Weixiang Shen received his BEng, MEng and PhD degrees in electrical engineering. Dr Shen is an Associate Professor at the Faculty of Science, Engineering and Technology, Swinburne University of Technology, Melbourne, Australia. From 1993 to 1994, he was a Visiting Scholar at The University of Stuttgart, Stuttgart, Germany, where he worked on battery management systems for photovoltaic systems. From 2003 to 2007, he was a Lecturer and Senior Lecturer at Monash University, Malaysia. From 2008 to 2009, he was a research fellow with Nanyang Technological University, Singapore, focusing on battery management systems for purifying wastewater based on membrane technology using solar photovoltaic systems. His research interests include battery charging, battery capacity estimation, battery management systems and integration of electric vehicles and renewable energy sources into power grids. He has authored or co-authored more than 80 papers published by peer-reviewed journals in the relevant research areas.

Dr Shen is an editor for the journal *Vehicles*, a guest editor for a special issue on "Advanced Energy Storage Techniques towards Sustainable Transportation" for the journal *Sustainability*, and a guest editor for a special issue on "Advanced Energy Storage Technologies and Their Applications" for the journal *IEEE Access*. He is an active reviewer for leading journals such as *IEEE Transaction on Vehicular Technology, IEEE Transaction on Power Electronics, IEEE Transaction on Industrial Electronics, IEEE Transaction on Energy Conversion,* and *Journal of Power Sources*. He also served as a conference organizing committee member and session chair for a few international conferences, such as *The IEEE International Conference on Industrial Electronics and Applications (ICIEA)* in 2008–2018. He was a general chair of the *International Conference on Energy, Ecology and Environment (ICEEE2018)* held in Melbourne, Australia, 2018.

Foreword by Professor Sun

The promotion of electric vehicles is part of a national strategy to reduce oil consumption and air pollution in many countries around the world. In China, the electric vehicle industry has been designated by the central government as a strategic emerging industry that represents what is considered to be the future of automobiles. It is well known that the advancement of battery technologies is crucial to the safe and efficient operation of electric vehicles. The two fundamental characteristics about battery technologies that affect the cost of operation, performance, and durability are power density and energy density. For vehicle applications, it is desirable that batteries have both high power density and high energy density, but there is generally a trade-off between these two characteristics, resulting in higher power density with a correspondingly lower energy density, or higher energy density with a lower power density.

In recent years, lithium-ion batteries have been widely accepted for electric vehicle applications due to their superiority in high energy density, high power density, and long cycle life. They exhibit strong coupling effects among electric, thermal and mechanical behaviors in electric vehicle applications, leading to strong time-varying, ambient temperature dependent and nonlinear characteristics. However, there are few measurable parameters for controlling and monitoring batteries. Currently, the most popular parameters are battery terminal voltage, charge/discharge current, and surface temperature. The lack of effective measurable parameters further complicates and challenges the development of lithium-ion battery management systems for electric vehicles. The key technologies involved in battery management systems include battery modeling, battery state estimation, battery charging and battery balancing. The fundamental solution to achieving active management of battery systems is to improve battery model accuracy, develop robust multi-scale and multi-state estimation approaches, and optimize charging and balancing processes. Research on any of these topics would contribute to the improvement of battery management systems that could reduce the risk of fire or explosion, caused by overcurrent, overvoltage, or overcharge/discharge.

The book *Advanced Battery Management Technologies for Electric Vehicles* is the culmination of more than a decade of research by Associate Professors Rui Xiong and Weixiang Shen on all important aspects of battery management systems for electric vehicles, including battery system modeling, state of charge estimation, state of energy estimation, state of health estimation, state of power estimation, battery charging and battery balancing, and the implementation of battery management systems. In particular, the book has a comprehensive coverage of the technical details of the core algorithms, which can realize the main functions of battery management systems in electric vehicles. Therefore, this book is not only a valuable reference for professionals, researchers and practicing engineers in battery management systems in electric vehicles and energy storage, but it can also be used as a course book for undergraduate as well as graduate students in engineering, particularly in automotive and electrical engineering.

Associate Professor Xiong is a former PhD student of mine. He started his master and doctoral programs at Beijing Institute of Technology (BIT) in 2008 and 2010, respectively. Immediately after receiving his PhD in 2014, he joined the research team in the National Engineering Laboratory for Electric Vehicles at BIT. Since then, he has been focusing on the research and development of battery management technologies in electric vehicles and carrying out systematic and in-depth investigations in cutting edge electric vehicle technologies on battery testing, battery modeling, battery states estimation, durability, safety and battery system integration and management, yielding fruitful achievements. This book is the representation of his persistent efforts in the development of advanced battery management technologies. It has also resulted from his close collaboration with Associate Professor Shen, an internationally renowned expert in battery state estimation, battery charging and battery balancing for electric vehicles.

I highly recommend this book not only because it is the first book exclusively devoted to advanced battery management technologies but also because it is the brainchild of Associate Professor Xiong who has made outstanding contributions to the development of battery management algorithms and has played a unique role in promoting advanced and intelligent battery management systems for all-climate electric vehicles. The results reported in this book are based on the technological achievements of the National Engineering Laboratory for Electric Vehicles at BIT and attributed to his extensive cooperation with top electric vehicle makers in China such as BAIC BJEV, ZhengZhou Yutong Bus, Huawei, and United Automotive Electronics. In summary, this book is a must-read for anyone who wants to understand the core algorithms and relevant technologies of battery management systems for electric vehicles.

Fengchun Sun
Professor and Academician, Beijing Institute of Technology
Director, National Engineering Laboratory for Electric Vehicles
Director, Collaborative Innovation Center of Electric Vehicles in Beijing
Director, National New Energy Vehicle Monitoring and Management Center

Foreword by Professor Ouyang

Two great challenges facing the Chinese automotive industry are climate change and energy security. First, China is the largest country in automotive production and sales in the world. Chinese vehicle sales and production exceeded 28 million in 2017, accounting for one-quarter of total world sales, and ranked first in the world for nine consecutive years. This has caused severe urban air pollution. Research shows that tailpipe emissions of current internal combustion engine vehicles are the main source of urban air pollution. They account for approximately 24% of pollution in some major cities, including Beijing, Tianjin, and Shanghai. Secondly, China's crude oil consumption is increasing greatly with the rapid growth of the ownership of internal combustion engine vehicles. China surpassed the United States for the first time to become the world's largest importer of crude oil in 2017. China's crude oil imports reached 8.43 million barrels a day in 2017, up 10% from 2016, compared with the US's 7.91 million barrels a day, leading to accelerating eastward movement of global oil trade.

These challenges require the development of new energy vehicles. The new energy vehicles change the propulsion system from engine to motor and thus essentially change energy sources from fossil fuels to electrochemical energy storage systems, where the stored energy can be derived from renewable energy sources such as wind or solar energy. Consequently, new energy vehicles can reduce urban air pollution and diversify energy sources. Furthermore, mass penetration of new energy vehicles can also lead to integrated sustainable transportation and power grids. Among all the new energy vehicles, electric vehicles such as pure electric vehicles and plug-in electric vehicles are becoming very attractive for road transportation. By the end of 2015, China had become the world's largest electric vehicle market, and the development of electric vehicles has been determined to be a national strategy for China.

Electric vehicles are partially or wholly driven by a battery system which consists of a combination of series and parallel connections of many battery cells. Lithium-ion battery cells are currently the most promising for the construction of battery systems due to their favorable performances in energy density, power density, energy efficiency, and life time. The battery system in electric vehicles experiences dynamic and complex operation conditions. Its performances vary strongly with many factors including battery temperature, charge and discharge rate, aging effect, depth of discharge and cell inconsistency in the battery pack for electric vehicles. Therefore, it is indispensable to develop advanced battery management technologies to monitor and control the battery system, thereby assuring its safe and reliable operation.

This book by Associate Professors Rui Xiong and Weixiang Shen presents their research results and contributions in advanced battery management technologies made over more than ten years. The book starts with the fundamental knowledge of battery electrochemistry and electric vehicle dynamics, driving cycles and requirements of battery management systems. It continues with the detailed provision of battery modeling techniques focusing on equivalent circuit models, model-based estimation methods for state of charge, state of energy, state of health and state of power, and battery charging and balancing techniques. These techniques are all critical in making a safer and more reliable battery system. Finally, the book ends with the integration of all these techniques into battery management systems for electric vehicles. Thus, this book covers the necessary background and techniques for the development of advanced battery management systems for electric vehicles.

A scientific national strategy for 2016–2020 is expected to play a critical role in making China the global leader in the electric vehicle industry. This book is published at a particularly timely moment when the electric vehicle industry in China is in the process of transforming from one that is investment driven to one that is innovation driven. This transformation requires new knowledge and innovative techniques in one of the key technologies for electric vehicles: battery management technologies.

I recommend this book not only because of its solid technical content but also because of the important and unique role Associate Professor Xiong plays in the systematic and original research work for the development of advanced battery management systems. I believe this book can benefit senior undergraduate and postgraduate students who are going to enter the electric vehicle industry. Chemical, mechanical and electrical engineers who are already in the electric vehicle industry can also benefit by systematically learning advanced battery management technologies from this book. Researchers who are working in academia can use this book as an in-depth source and comprehensive reference to develop new battery management technologies for electric vehicles.

Minggao Ouyang
Professor and Academician, Tsinghua University
Director, State Key Laboratory of Automotive Safety and Energy
Director, China–US Clean Energy Research Centre-Clean Vehicle Consortium

Series Preface

Batteries have been used in vehicles for well over a century, and with the advent and rise of electric vehicle, the battery is arguably the single most important element of the electric vehicle. As newer generations of electric vehicles including hybrid, plug-in, and extended range vehicles become available and more popular, the impact of fully exploiting the battery during its entire lifecycle is growing in significance. Squeezing the best performance out of a battery and maximizing its useful life are not only important to electric vehicle OEMs and consumers, but they are key to ensuring that future electric vehicle fleets are highly sustainable.

The *Automotive Series* publishes practical and topical books for researchers and practitioners in industry, and postgraduate/advanced undergraduates in automotive engineering. The series covers a wide range of topics, including design, manufacture and operation, and the intention is to provide a source of relevant information that will be of interest and benefit to people working in the field of automotive engineering. *Advanced Battery Management Technologies for Electric Vehicles* is an excellent addition to the series focusing on optimizing the performance and extending the life of one of the most critical elements of any electric vehicle, the battery. While a significant amount of literature available to the modern automotive engineer focuses on concepts such as battery design and specifications, there is a lack of information regarding the management of the battery including the battery's health, charging cycles, and performance. It is the only text currently available that discusses practical implementation of use strategies for batteries that are important to the battery's, and thus the vehicle's, performance. It also presents key concepts that are related to the overall life cycle of the battery. This wealth of information is not only critical to the performance of the electric vehicle, but it is also paramount to the value proposition of the vehicle. For example, choosing the correct charging and discharging cycles for the battery will have a direct effect on the life span and performance of the battery, directly affecting the marketability and sustainability of the vehicle.

As is mentioned in the beginning of this preface, *Advanced Battery Management Technologies for Electric Vehicles* is part of the *Automotive Series*; however, batteries are found on a wide variety of other systems outside of the automotive sector. Thus, the concepts presented in this text are applicable across a wide variety of fields. In particular, most of our next generation renewable energy sources require the ability to store energy for use at a later time. For example, solar plants generate significant power during the day, but not at night. Energy storage units employing large battery banks are one means by which solar energy may be tapped when the sun goes down. Issues related to battery management technologies such as charging, balancing, charge and state estimation will be valuable in any technology sector that employs rechargeable batteries for high power applications. This makes the content of the text applicable across a wide variety of technology fields, and of significant use beyond the automotive sector.

Advanced Battery Management Technologies for Electric Vehicles provides a set of well-focused and integrated topics that are critical to battery systems management. It presents these topics with a special focus on the automobile. The text employs a set of well thought out case studies, clearly demonstrating the utility and application of the fundamental concepts that are developed by the authors. It is state-of-the-art text, written by recognized experts in the field and is a valuable resource for practitioners in the field. It is an excellent addition to the *Automotive Series*.

Thomas Kurfess
November 2018

Preface

Electric vehicles (EVs) have been widely recognized as the most environmentally friendly form of road transport. Over the past decade, there have been significant advancements in EV technologies. Such advancements have seen EVs gradually replace conventional internal combustion engine vehicles (ICEVs). Some experts foresee sale volumes of EVs will surpass those of ICEVs in the next 10–20 years. In the current global EV market, China has played a dominant role in EV manufacturing and sales. China's EV sales have topped the world for the third consecutive year since 2015 and are predicted to account for 57% of the world's EV sales by 2035.

With the rapid uptick in EV sales, EV technologies have been developed at an accelerated pace. Among these EV technologies, researchers have been conducting studies on battery management technologies for many years. Weixiang Shen is a leader of applied battery research for EVs and renewable energy systems at Swinburne University of Technology. He is a pioneer researcher on battery management technologies for EVs and has been working in the area for more than 20 years. Rui Xiong is a leader of advanced energy storage and application at Beijing Institute of Technology. He has been working on battery management technologies for about 10 years. They both have published numerous conference and journal papers, successfully completed many industrial projects and engaged in consultancy work on the topic of battery management technologies for EVs. While conducting research, the authors have found that although there is a wealth of information for battery management technologies in the public domain literature, there is not a comprehensive and specific book focusing on battery management technologies in EVs as yet. The aim of this book is to bridge this gap.

Lithium-ion (Li-ion) batteries have been widely used in EVs due to their high energy density, high power density, high voltage, low self-discharge, and long cycle life in comparison with other secondary batteries. However, behaviors of Li-ion batteries are greatly affected by their working environment. In particular, Li-ion battery systems in EVs operate in a more dynamic working environment than those in portable electronic devices such as laptops and mobile phones. The charge and discharge currents and thus voltages of the battery systems fluctuate significantly when EVs are in regenerative braking and acceleration and their operation temperatures vary greatly when EVs are driving during different seasons in various locations. The battery systems under such large current and temperature variations along with rapid charge–discharge cycles require sophisticated battery management systems (BMSs). The purpose of BMSs is to regulate the operation of the battery systems within allowable voltage, current and temperature ranges and to estimate battery states for EV optimal operations. This leads to the development of advanced battery management technologies, which will be presented in this book.

To make this book self-contained, vehicle dynamics and standard EV driving cycles are introduced to provide the basis to discuss power and energy requirements of EVs and to evaluate EV performances. Also included is an introduction to the electrochemistry of different battery systems applicable to EVs. The key battery management technologies and BMSs have also been introduced. Beyond the basics, the book focuses on battery modeling technique and estimation techniques of battery state of charge, state of energy, state of health and state of power. For each of these techniques, there is the detailed description of the techniques accompanied with step-by-step mathematical equations. Case studies are provided with experimental and simulation results to demonstrate the applications of these techniques in EV working conditions. Furthermore, this book discusses battery charging and balancing techniques. The application of all the above-mentioned techniques to BMSs and the key technologies of BMSs in future generation are also discussed. In addition to referencing the relevant work of other researchers, a large portion of the materials presented in this book is the collection of many years of research and development by the authors.

This book consists of eight chapters. In Chapter 1, the fundamental knowledge about EVs, requirements of battery systems for EVs and battery systems applicable to EVs such as lead–acid, nickel–cadmium, nickel–metal hydride and Li-ion batteries are introduced. The overview of the key battery management technologies, the BMS structures and BMS functions are presented.

In Chapter 2, the classification of battery modeling techniques is provided. Then, the chapter goes on to focus on the explanation of battery equivalent circuit models in terms of the model structure, open circuit voltage, polarization and hysteresis characteristics. Based on the equivalent circuit models, offline parameter identification methods and online parameter identification methods are introduced, followed by case studies showing the application of each method. Furthermore, the influences of battery aging, battery type and battery temperature on accuracies of equivalent circuit models are discussed in detail.

In Chapter 3, the classification of battery state of charge estimation methods is introduced. This includes look-up table methods, an ampere-hour integral method, data-driven estimation methods, and model-based estimation methods. Following this, a detailed explanation is given to model-based state of charge estimation methods with constant model parameters using Kalman filter and an H infinity filter. The influences of the uncertainties on the state of charge estimation methods are discussed. For real EV applications, model-based state of charge and state of energy estimation methods with identified model parameters in real-time are emphasized. The MATLAB codes and Simulink models used in the case studies to implement these estimation methods are provided to users.

In Chapter 4, battery state of health estimation is divided into two categories: experimental methods; and model-based methods. The experimental methods include direct measurement methods and indirect analysis methods. The model-based methods include adaptive state estimation methods and data-driven methods. For real EV applications, the focus is on model-based methods including the joint estimation method and dual estimation method which allow the estimation of state of health together with state of charge. Readers should be able to follow the examples of MATLAB codes and Simulink models to design a state of health estimator for their EV applications.

In Chapter 5, battery state of power estimation has been grouped into instantaneous state of power estimation and continuous state of power estimation. The instantaneous state of power estimation includes the hybrid pulse power characterization method, the state of charge-limited method, voltage-limited methods and multi-constraint dynamic methods. For realistic state of power estimation in EVs, this chapter mainly addresses the continuous state of power estimation methods. Again, examples of MATLAB codes and Simulink models are provided for readers to implement these estimation methods.

In Chapter 6, basic terms are defined to evaluate the performances of charging methods. With these basic terms, different charging methods based on pre-set charging profiles and charging termination methods for Li-ion batteries are reviewed. The qualitative comparison of different charging methods for Li-ion batteries is also discussed. Subsequently, the latest development of charging methods is introduced in terms of two directions. One direction is the optimization of the charging profile during the charging process to minimize charging time or maximize charging efficiency or achieve the balance between charging time and charging efficiency. The other direction is the development of a new battery technology (i.e. lithium titanate oxide battery) to allow for high charge acceptance, speeding up the charging process.

In Chapter 7, several battery balancing techniques are introduced, including battery sorting, battery passive balancing and battery active balancing. Then, this chapter focuses on battery active balancing through detailed discussions of battery balancing criterion, balancing control and balancing circuits. Two examples of active battery balancing systems are provided to demonstrate their applications in EVs.

In Chapter 8, basic BMS functions are introduced. Thereafter, the typical BMS structures and representative BMS products are discussed. The hardware examples of BMS implementation are provided to demonstrate the applications of battery management technologies in EVs. This chapter also covers the discussion of key insights into future generations of BMSs including self-heating and safety management as well as the application of cloud computing and big data in the state of health estimation and battery life prediction.

The material in this book is recommended for a graduate or senior-level undergraduate course. Depending on the background of the students in different disciplines, course instructors have the flexibility to choose those chapters from the book that are most suitable for their students. This book is also an in-depth source and comprehensive reference in advanced battery management technologies for engineers and researchers working in EV-related industries, government organizations, and academia.

For the convenience of readers, MATLAB codes and Simulink models for the case studies are provided in this book. Readers can download them via the following link https://sites.google.com/view/weixiang-shen or contact the authors (Weixiang Shen at wshen@swin.edu.au or Rui Xiong at rxiong@bit.edu.cn) to arrange to receive a copy.

Despite the authors' best intentions, there may be some errors and confusing statements in this book. The authors welcome any suggestions and comments on this first edition of their book and will post any necessary corrections or amendments in a timely fashion on the website (https://sites.google.com/view/weixiang-shen), and ensure that any necessary corrections or amendments are incorporated in future editions.

Acknowledgments

The authors wish to acknowledge and thank all those who helped to complete the book, particularly the staff and graduate students from both the advanced energy storage and application at Beijing Institute of Technology and the applied battery research for EVs and renewable energy systems at Swinburne University of Technology. The staff to be acknowledged and thanked are Prof. Fengchun Sun, Prof. Hongwen He, Postdoctoral fellow Zeyu Ma, Dr Hao Mu, and Dr Fengxian He. The graduate students to be acknowledged and thanked are Cheng Chen, Jiahuan, Lu, Jinpeng Tian, Ju Wang, Ruixin Yang, Quanqing Yu, Yongzhi Zhang, Jiayi Cao, Yanzhou Duan, Yu Fang, Xinggang Li, Linlin Li, Zhirun Li, Liang Lü, and Kan Wang.

Sincere acknowledgments are included where appropriate. The authors have tried their best to cite work in the references listed in each chapter. If any sources have been missed, the authors apologize for the oversight and will rectify these omissions in future editions of the book. The names of any products or suppliers for the laboratory equipment used in this book are provided for information only and are not in any way to be perceived as an endorsement of such products or suppliers by the publisher or the authors.

Finally, the authors are extremely grateful to the publisher Wiley and the editorial staff for giving them the opportunity to publish this book and for helping in every way possible in the initiation of this book project and during the development and production of the manuscript.

Rui Xiong
Weixiang Shen
Melbourne
August 2018

1

Introduction

1.1 Background

With ever growing concerns about energy sustainability and environmental issues, the adoption of electric vehicles (EVs) such as pure electric vehicles (PEVs) and plug-in electric vehicles (PHEVs) has been identified to be one of the most effective strategies to reduce dependency on fossil fuels and greenhouse gas emission, and has attracted more and more attention from governments, industries, and customers [1, 2]. Many countries have taken aggressive step to promote EVs to meet emission targets under the Paris climate accord [3]. France has announced its plan to stop selling petrol and diesel-powered cars by 2040. The UK follows France in banning sales of new petrol and diesel cars from 2040. Norway, which has the highest number of EVs in the world, has set a target of only allowing sales of 100% PEVs or PHEVs by 2025. The Netherlands has mooted a 2025 ban for diesel and petrol cars. Germany has passed a resolution calling for a ban on combustion engine cars by 2030. India is mulling the idea of moving toward prohibiting the internal combustion engine in 2030. China is developing a plan to phase out vehicles powered by fossil fuels [4]. Though China has not yet suggested any concrete timeline, it has ambitious goals for automotive efficiency and climate change including a cap on carbon emissions by 2030. Experts suggest this new ban might come into force around then. The policies of restricting sales of diesel and petrol cars in these countries have stimulated carmakers around the world to accelerate the research and development of EV technologies and the commercialization of EVs.

EVs in this book refer to vehicles which are wholly or partially driven by an electric motor using energy stored in batteries. They have merits over internal combustion engine vehicles (ICEVs) in terms of the energy consumed, which can be generated from various energy sources including renewable energy sources (RESs). As the energy of batteries is depleted, they are recharged by the electricity from power grids. In order for EVs to be truly clean and sustainable road transportation, electricity generation needs to come from power grids with the integration of more RESs [5]. However, the intermittent nature of RESs adversely affects power grid voltage, frequency, and reactive power. Such a power grid needs to be compensated by energy storage which stores the excessive power generated from RESs and delivers power back into grids to offset high demand during peak hours [6]. The batteries in EVs are good candidates as energy storage and they can play the role of power compensation which enhances power grid resilience and provides an opportunity for better acceptance of RESs. As such, the batteries in an EV not only provide energy to drive it but also exchange energy with power grids to stabilize

Advanced Battery Management Technologies for Electric Vehicles, First Edition. Rui Xiong and Weixiang Shen.
© 2019 John Wiley & Sons Ltd. Published 2019 by John Wiley & Sons Ltd.

power systems. In such applications, the batteries operate at dynamic and deep cyclic conditions that can lead to poor performances and premature aging [7, 8]. Advanced battery management technologies (BMTs) are needed to monitor and control the batteries in EVs for performance improvement and life extension. BMTs involve battery modeling, battery states estimation, battery charging, battery balancing, and battery thermal management [9], which will be discussed in the rest of the book. In this chapter, the fundamentals of EVs are first explained. Then, energy and power requirements of batteries in EVs and performances of current battery technologies are discussed. Finally, the key BMTs and the functions of battery management systems (BMSs) will be briefly introduced.

1.2   Electric Vehicle Fundamentals

Depending on the configuration of an EV, the propulsion power and energy of the EV can be partially or wholly supplied by the batteries installed in it. Without loss of generality, a PEV is used as an example to discuss the working principle. Similar to ICEVs, the powertrain in a PEV provides power to move a vehicle and energy to cover a certain driving distance under different road conditions and driving modes [10, 11]. The force that a vehicle must overcome to travel is known as road load. As shown in Figure 1.1, the road load F_L consists of three major components: aerodynamic drag force F_d, rolling resistance force F_r, and climbing force F_c as given by

$$F_L = F_d + F_r + F_c \tag{1.1}$$

The aerodynamic drag force is the friction of the vehicle moving through the surrounding air and is expressed as

$$F_d = \rho A C_d (V - V_w)^2 / 2 \tag{1.2}$$

where ρ is the density of the surrounding air, A is the vehicle frontal area, C_d is the aerodynamic drag coefficient, V is the vehicle speed, V_w is the wind speed in the opposite direction of the moving vehicle, and θ is the angle of the slope.

The rolling resistance force is caused by the deformation of the wheel and road surface. The deformation of the wheel heavily dominates the rolling resistance while the

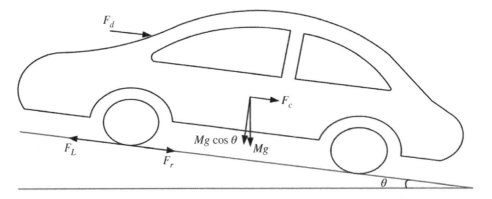

Figure 1.1 Forces applied on a vehicle.

deformation of the road surface is generally insignificant. The rolling resistance force is normally calculated by

$$F_r = MgC_{rr} \cos \theta \qquad (1.3)$$

where M is the overall mass of the vehicle, g is the gravitational acceleration, and C_{rr} is the rolling resistance coefficient.

The climbing force is the downward force for a vehicle to climb up a slope which is the resolution of the gravity force due to the vehicle weight that acts along the slope

$$F_c = Mg \sin \theta \qquad (1.4)$$

According to Newton's second law, the acceleration of a vehicle is determined by all forces applied on it, which is given by

$$a = (F_t - F_L)/(k_t M) \qquad (1.5)$$

where F_t is the total traction force available to overcome the road load F_L and drive the vehicle with the acceleration a, and k_t is the mass factor that converts the rotational inertia of rotating components into equivalent translational mass. As a result, the total propulsion force can be expressed as

$$F_t = k_t M\alpha + F_d + F_r + F_c \qquad (1.6)$$

The power demand to drive the vehicle at the speed V can be calculated by

$$P_t = F_t V = k_t M\alpha V + F_d V + F_r V + F_c V \qquad (1.7)$$

Correspondingly, the energy required to move the vehicle for a certain distance from time t_0 to t can be calculated by

$$E_t = \int_{t_0}^{t} P_t dt \qquad (1.8)$$

Depending on the traction force F_t, the vehicle can operate at three driving modes:

- *Traction mode.* $F_t > 0$, the traction force overcomes road load to move the vehicle forward.
- *Braking mode.* $F_t < 0$, the kinetic energy of the vehicle is dissipated by brakes or regenerative braking, and the propulsion system can either work or be cut off.
- *Coasting mode.* $F_t = 0$, the vehicle propulsion system stops working, and all the kinetic energy is deceased and transformed to the resistance losses.

It should be noted that EV batteries only need to provide energy in the traction mode.

1.3 Requirements for Battery Systems in Electric Vehicles

Three common parameters are normally used to assess EV performance [10]:

- *Range per charge.* This is the driving distance when the batteries in EVs are fully charged. It may vary with the conditions of use and can be determined on the basis of an operation condition of a constant speed on a level road or a type of driving cycle or the loosely defined urban or highway situations. For example, an EV can offer 250 km per charge at a constant speed of 60 km h^{-1} or 170 km per charge under a new European driving cycle (NEDC) combined cycle.

- *Acceleration rate.* This is usually represented by the minimum time required to accelerate the vehicle from zero to specific speeds, such as 40, 60, or 80 km h^{-1}.
- *Maximum speed.* This is the maximum safe speed which the vehicle can attain. It does not mean the speed limit on the public road but mainly indicates vehicle performance.

1.3.1 Range Per Charge

Standard driving cycles are generally used to evaluate the driving range for EVs in different regions and countries [12, 13].

(1) *In the USA.* The Urban Dynamometer Driving Schedule (UDDS) is the most common driving cycle, which is also known as the FTP-72 cycle or LA-4 cycle or Federal Urban Driving Schedule (FUDS). It was developed originally to evaluate the noxious emissions of ICEVs and was based on a cycle derived from the statistical flow of traffic patterns in Los Angeles. Subsequently, it has been widely used to evaluate the fuel economy of urban or city driving. In contrast, the Federal Highway Driving Schedule (FHDS) was developed to represent driving patterns in rural or cross-country conditions. It is also known as The United States Environmental Protection Agency (US EPA) Highway Fuel Economy Test (HWFET) or US EPA HWFET. Figures 1.2 and 1.3 show their corresponding speed–time profiles.

(2) *In Europe.* The ECE-15 cycle was the first driving cycle to be legislated in the European Union. The name ECE-15 corresponds to United Nations Economic Commission for Europe (UNECE) Regulation No. 15 published on April 11, 1969. It was developed to represent the urban driving cycle (UDC), e.g. in Paris or Rome, as shown in Figure 1.4. It is characterized by low vehicle speeds, low engine load, and low exhaust gas temperature. The Extra Urban driving cycle (EUDC) was

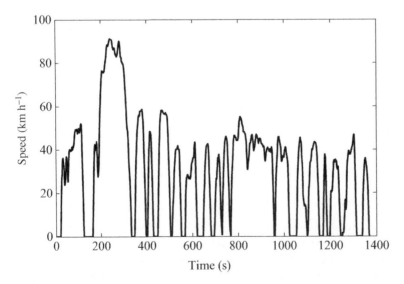

Figure 1.2 Speed versus time under UDDS.

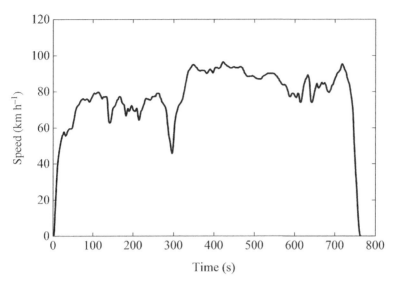

Figure 1.3 Speed versus time under FHDS.

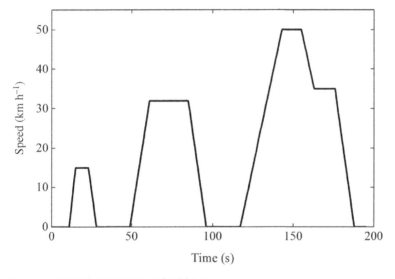

Figure 1.4 Speed versus time under ECE-15.

developed to include more aggressive and high speed driving modes, as shown in Figure 1.5. NEDC is a composite of four repeating ECE-15 cycles and one EUDC, as shown in Figure 1.6. NEDC is widely adopted in Europe to evaluate fuel economy and emissions of ICEVs. It is also adopted to evaluate the driving range and energy consumption of EVs.

(3) *In Japan.* Japanese 10.15 Mode was developed to reflect metropolitan driving conditions as well as highway driving conditions, based on driving performance studies in major cities in Japan. The profile of the Japanese 10.15 Mode is composed of three

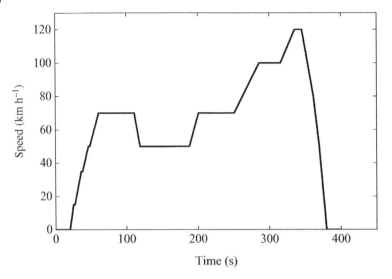

Figure 1.5 Speed versus time under EUDC.

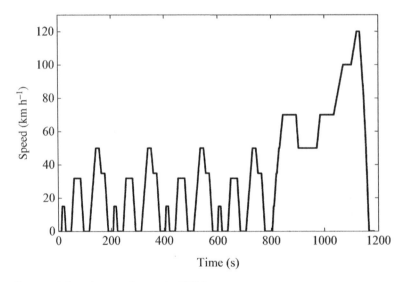

Figure 1.6 Speed versus time under NEDC.

repetitions of J10 followed by J15, as shown in Figure 1.7. As a regulation, the driving range of EVs in Japan is evaluated only by the Japanese 10.15 Mode driving cycle.

To evaluate power and energy requirements for the batteries in EVs under the above standard driving cycles, a typical PEV with the parameters shown in Table 1.1 is taken as an example [5].

When the typical PEV is running in these standard driving cycles, the batteries in it will supply power and energy to vehicle movement as well as auxiliary units. For

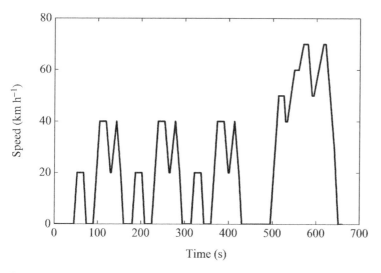

Figure 1.7 Speed versus time under Japanese 10.15 Mode.

Table 1.1 Parameters of a typical PEV.

Mass, M	1360 kg
Mass factor, k_t	1.05
Acceleration rate, 0 to 100 km h^{-1} in 10 s	2.68 m s^{-2}
Coefficient of rolling resistance, C_{rr}	0.01
Air density, ρ	1.225 kg m^{-3}
Vehicle frontal area, A	2 m^2
Aerodynamic drag coefficient, C_d	0.5
Wind speed, V_w	0 m s^{-1}
Road slope angle, θ	0°
Efficiency of gearbox to wheels	1.0
Efficiency of motor to gearbox	0.9
Efficiency of inverter to motor	0.95
Efficiency of batteries to inverter	0.95

simplicity, this book only discusses the energy and power demands of vehicle kinetics and ignores those required by auxiliary units. The five standard driving cycles of UDDS, FHDS, ECE-15, NEDC, and Japanese 10.15 Mode are used to analyze the power and energy demands of this PEV. According to Eq. (1.7), the power demands corresponding to the five standard driving cycles are calculated and their results are shown in Figures 1.8–1.12, respectively.

Since the batteries provide energy to drive the EV only in the traction mode, the energy demands in the traction mode are calculated under the five standard driving cycles by

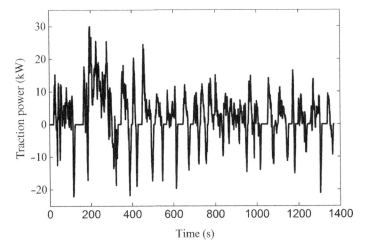

Figure 1.8 Traction power versus time under UDDS.

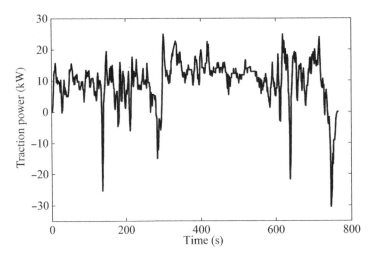

Figure 1.9 Traction power versus time under FHDS.

using Eq. (1.8), and the corresponding results are shown in Table 1.2. In current EV markets, the average size of the battery systems in EVs is around 30 kWh [14, 15]. Thus, it is assumed that the typical PEV has a 30 kWh battery system installed for the evaluation of the range per charge running at the repetitions of the five standard driving cycles until the battery system discharges to 80% of its total energy (i.e. 24 kWh). The corresponding driving distances are shown in Table 1.3.

Generally, an ICEV can easily give a driving range of 480–640 km (i.e. 300–400 miles) with a full tank of gasoline (e.g. 16 gal). For the same driving range, the PEV would require a battery system larger than 100 kWh that weighs over 667 kg, which is too big and heavy. This is due to the fact that gasoline has a theoretical specific energy of

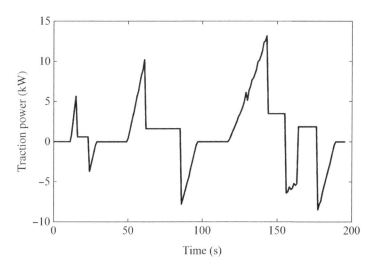

Figure 1.10 Traction power versus time under ECE-15.

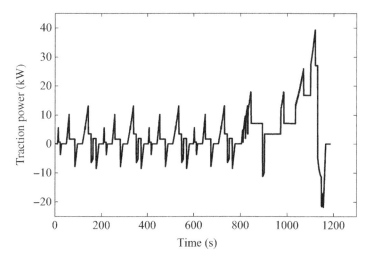

Figure 1.11 Traction power versus time under NEDC.

13 000 Wh kg^{-1}, which is about 100 times higher than the specific energy of 150 Wh kg^{-1} of typical lithium-ion (Li-ion) batteries [16]. However, since the efficiency of a propulsion system in EV powertrains is much higher than that in ICEVs, less energy is needed to propel EVs. Considering the efficiency of the propulsion system is 80% for EVs and 20% for ICEVs, the total amount of energy stored for an EV can be a quarter of what an ICEV needs for the same mileage. Thus, it is realistic to have a battery system of around 30 kWh to achieve around a 200 km (i.e. 124 miles) range based on current battery technologies, where the batteries in EVs are only allowed to discharge to 80% of the total energy (i.e. 24 kWh).

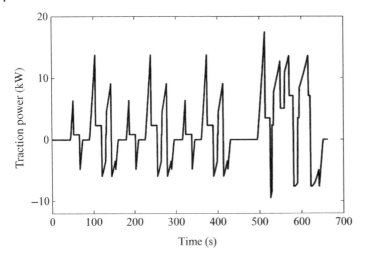

Figure 1.12 Traction power versus time under Japanese 10.15 Mode.

Table 1.2 Energy demands for five standard driving cycles.

	UDDS	FHDS	ECE-15	NEDC	Japanese 10.15
Driving distances (km) per standard driving cycle	11.99	16.51	0.99	10.93	4.17
Energy demands in traction mode, kJ (kWh)	5132 (1.42)	7923 (2.20)	602 (0.17)	5024 (1.40)	1665 (0.46)

Table 1.3 Driving distances for repetition of five standard driving cycles.

	UDDS	FHDS	ECE-15	NEDC	Japanese 10.15
Driving distances without regenerative braking (km)	202	180	140	187	218

1.3.2 Acceleration Rate

The vehicle acceleration is driven by all forces applied on it. This force can be expressed as

$$F_A = M\alpha \tag{1.9}$$

where $\alpha = dV/dt$ is the vehicle acceleration magnitude. The energy required for vehicle acceleration from standstill to a reference speed V_r is calculated by

$$E_A = MV_r^2/2 \tag{1.10}$$

EV batteries provide the energy to accelerate the PEV. If the time t is given for the acceleration process, the relationship between the energy and mean power can be represented approximately by

$$E_A \approx \overline{P}t \tag{1.11}$$

Considering the power increases from zero to the maximum power P_{max} in the acceleration process, the approximate mean power is

$$\overline{P} \approx P_{max}/2 \tag{1.12}$$

Substituting Eqs. (1.11) and (1.12) into Eq. (1.10) leads to

$$P_{max} \approx MV_r^2/t \tag{1.13}$$

If the acceleration rate is expected to be $100\,km\,h^{-1}$ within 10 seconds for the typical PEV, the maximum power which the batteries need to provide is $104\,kW$ approximately to the discharge rate of 3.5C for the 30 kWh battery system.

1.3.3 Maximum Speed

The maximum speed is an important parameter to indicate EV performance. Since the aerodynamic drag force is the cube of speed, it becomes the dominant factor at high speeds compared with other resistance forces. Thus, the maximum power needed to calculate the maximum speed V_{max} on a level road with ideal gearing can be simply described as

$$V_{max} = \sqrt[3]{2P_{max}/(\rho A C_d)} + V_w \tag{1.14}$$

If the maximum power delivered from the battery system with 30 kWh in the typical PEV is 120 kW which is equivalent to a discharge rate of 4C, then the corresponding maximum speed is found to be $209\,km\,h^{-1}$ by substituting 120 kW and the parameters in Table 1.1 into Eq. (1.14).

1.4 Battery Systems

Batteries are normally categorized into two types, namely primary and secondary batteries. Primary batteries have non-rechargeable properties and can only be discharged once. These types of batteries are also known as "dry cells." Secondary batteries have rechargeable properties and can be recharged to their original conditions after discharging. These types of batteries are also known as rechargeable batteries. Battery systems in EVs or PHEVs should provide rechargeable properties with satisfactory performance, including high energy density for covering sufficient driving range, high power density for a vehicle to achieve desired acceleration rate and maximum speed, a long cycle life for minimum maintenance and low cost for better market acceptance. Secondary batteries fulfill most of the above-mentioned criteria and thus are widely used in EVs.

The development of secondary batteries has continued for over 100 years. So far, many different types of secondary batteries have been available in commercial markets. Some of them have been widely used in EVs as energy storage; these are lead–acid, nickel–cadmium (NiCd), nickel–metal hydride (NiMH) and Li-ion batteries [10]. EVs require battery systems to store sufficient energy to meet desired driving distances and provide adequate peak power to deliver a specific acceleration performance. To better understand battery systems in EVs, we first provide an introduction to the electrochemistry of secondary battery cells as the basic building blocks for making battery systems. Then, we present the backgrounds, operating principles, performances and applications of the above-mentioned four types of secondary batteries.

1.4.1 Introduction to Electrochemistry of Battery Cells

Battery cells store energy based on the conversion of chemical energy into electrical energy and vice versa through an electrochemical reaction [16–18]. Figure 1.13 shows a schematic diagram of an electrochemical battery cell. The cell is made up of a negative electrode, a positive electrode, an electrolyte, a separator, a positive current collector, and a negative current collector, and it supplies power to an external load. The solid and porous electrodes allow the electrolyte to pass through. The separator prevents electrons from flowing but allows positive and negative ions to migrate between the two electrodes through the electrolyte. The positive and negative current collectors provide a pathway for electrons to flow through an external load. The cell makes use of the fact that different materials in the positive and negative electrodes immersed in an electrolyte solution have different tendencies to gain and lose electrons.

These electrons will flow through the external load while the resultant ions move in the electrolyte solution, forming a closed circuit to create the current and providing power and energy to the external load. It is convenient to introduce the terms "anode" and "cathode." A cathode is defined as the electrode which accepts electrons from the external circuit or as the electrode at which reduction reaction takes place. Conversely, an anode is defined as the electrode which releases electrons to the external circuit, or as the electrode at which oxidation reaction takes place. The basic working principle of a battery cell is explained below.

During discharge, the positive electrode is the cathode and the negative electrode is the anode. Positive ions flow from the anode to the cathode through the electrolyte and separator. Negative ions flow in the opposite direction. Electrons flow from the anode (negative electrode) to the cathode (positive electrode) through the external load. The anode builds up negative charge and the cathode builds up positive charge, generating a cell voltage V_t. When this voltage is applied across the external load, a current I_t will be produced in the opposite direction of electron flow.

During charge, the process is reversed. The positive electrode is the anode and the negative electrode is the cathode. The positive ions flow from the anode to the cathode through the electrolyte and separator. Negative ions flow in the opposite direction. Electrons flow from the anode (positive electrode) to the cathode (negative electrode) through the external load. The negative electrode material dissolves in the electrolyte solution to form negative ions and gains an electron in what is called the reduction reaction. The positive electrode loses an electron by depositing positive ions from the electrolyte in what is called an oxidation reaction. These oxidation–reduction reactions are reversible in secondary batteries. Thus, batteries can be discharged to return the electrodes to their pre-charged states or charged to their pre-discharged states.

The ions move through the electrolyte under diffusion and migration. Diffusion results from the existence of a concentration gradient in the electrolyte. Over time, if there is no ion production, the ions in the electrolyte diffuse evenly throughout the cell. Migration results from the presence of the electric field generated by the positive and negative

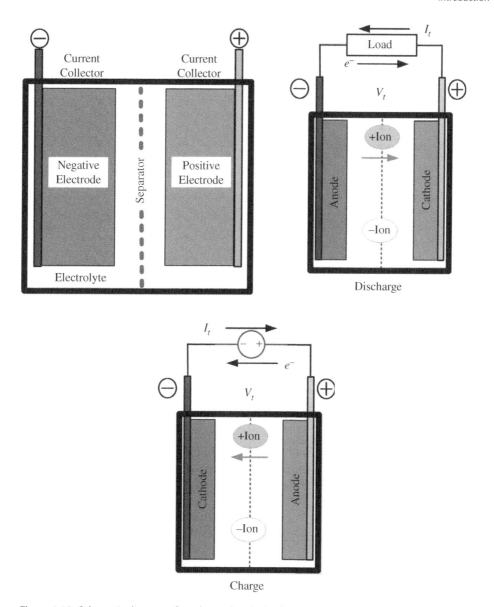

Figure 1.13 Schematic diagram of an electrochemical cell.

electrodes. The positive ions migrate toward the negative electrode and the negative ions migrate toward the positive electrode. The movement of ions through the electrolyte and electrons through the external circuit enables the storage and release of energy.

When the current passes the cell, the cell terminal voltage V_t deviates from its equilibrium state voltage V_{eq}; such deviation is called a polarization voltage or overvoltage η. The polarization voltage incudes three major components:

- Ohmic overvoltage drop which is caused by the resistances in the bulk of the electrolyte, separators, electrodes and current collectors and other connectors.

- Activation overvoltage which is related to the charge transfer at each electrode/electrolyte interface and is also known as "electrode losses."
- Concentration overvoltage which is connected to the depletion or accumulation of active materials near the electrode surface.

In practical batteries, cell voltage for discharging can be expressed as

$$V_t^{dis} = E_{eq} - \eta^{dis} \tag{1.15}$$

while cell voltage undergoes charging

$$V_t^{cha} = E_{eq} + \eta^{cha} \tag{1.16}$$

where η^{dis} and η^{cha} represent the polarization voltages for charging and discharging processes, respectively.

1.4.1.1 Ohmic Overvoltage Drop

The source of ohmic overvoltage drop is the internal resistance of the bulk phases within the cell. If the current distribution is uniform, then for a phase with conductance σ, the resistance is

$$R = x/A\sigma \tag{1.17}$$

where x is the thickness of the phase and A is its cross-sectional area. For the passage of a current I_t through a cell with j sequential phases, Ohmic overvoltage drop is expressed by

$$\eta_{ohmic} = I_t \sum_j (x_j/A_j\sigma_j) \tag{1.18}$$

1.4.1.2 Activation Overvoltage

Activation overvoltage is associated with an electrode process which mainly involves two consecutive steps: (i) the transport of the electroactive species from the bulk of the electrolyte phase to the electrode surface; and (ii) interfacial charge transfer. For relatively low currents, it can be assumed that the surface concentration does not deviate significantly from the bulk value. Hence, the transfer of electrons or ions is the rate-determining step in the process which is directly controlled by potential difference or overvoltage across an electrode/electrolyte interface. The Butler–Volmer equation describes the relationship between electrical current I_t representing the rate of charge transfer and the activation overvoltage η_a as

$$I_t = i_o \left\{ \exp\left(\frac{-\alpha nF\eta_a}{RT} \right) - \exp\left[\frac{-(1-\alpha)nF\eta_a}{RT} \right] \right\} \tag{1.19}$$

where i_o denotes exchange current, α represents the transfer coefficient whose value is a constant between 0 and 1 (typically, it is close to 0.5), F is the Faraday constant, R is the universal gas constant, T is absolute temperature, and n is the number of electrons involved in the electrode reaction.

1.4.1.3 Concentration Overvoltage

For relatively large currents, the transport of electroactive species to the electrode surface becomes the limiting factor to restrict current. Concentration overvoltage η_c is used to describe such current restriction caused by a concentration gradient at the electrode surface. Assuming that diffusion is the only mechanism of transport, the concentration overvoltage can be evaluated using the relationship

$$\eta_c = \frac{RT}{nF} \ln \left(\frac{i_{\lim} - I_t}{i_{\lim}} \right) \tag{1.20}$$

where $i_{\lim}$ is a maximum or limiting current corresponding to the maximum concentration gradient.

1.4.2 Lead–Acid Batteries

Lead–acid batteries are the oldest type of secondary batteries and they are used extensively in communication systems, portable devices, heavy industry and EVs as well as stationary applications ranging from small emergency supplies to load leveling in power systems. The initial design of lead–acid batteries is the flooded versions which consist of a negative electrode of porous lead (Pb, lead sponge) and a positive electrode of lead oxide (PbO_2), both immersed in an aqueous solution of sulfuric acid (H_2SO_4). A microporous material as the insulator prevents the two electrodes from touching while allowing the electrolyte and the ions into it to enable conduction. A schematic diagram of a lead–acid battery is shown in Figure 1.14.

The chemical reaction at the positive electrode is

$$PbO_2 + 3H^+ + HSO_4^- + 2e^- \leftrightarrow PbSO_4 + 2H_2O \tag{1.21}$$

and that at the negative electrode is

$$Pb + HSO_4^- \leftrightarrow PbSO_4 + H^+ + 2e^- \tag{1.22}$$

Thus, the overall chemical reaction is

$$Pb + PbO_2 + H_2SO_4 \underset{charge}{\overset{discharge}{\rightleftharpoons}} 2PbSO_4 + 2H_2O \tag{1.23}$$

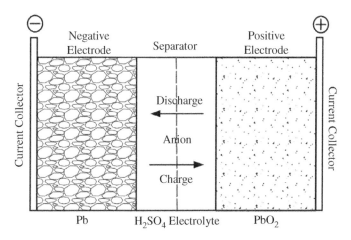

Figure 1.14 Schematic diagram of a lead–acid battery.

According to Eq. (1.23), when the battery is discharged, both electrodes turn into lead sulfate ($PbSO_4$) and the electrolyte loses its sulfuric acid (H_2SO_4) and eventually becomes water. When the battery is charged, the lead sulfate is converted into lead dioxide (PbO_2) at the positive electrode and the sponge lead at the negative electrode, and the electrolyte gains its sulfuric acid. The concentration of the electrolyte increases during the charging process and decreases during the discharging process, thus it can be used to determine the battery's state of charge (SOC).

The valve regulated lead–acid (VRLA) batteries are designed to promote the chemical recombination of the oxygen at the negative electrode to minimize water loss, so they do not require regular checking of the electrolyte level like the flooded lead–acid batteries. VRLA batteries have two forms of immobilized electrolyte: the absorbed electrolyte and the gelled electrolyte in a porous glass separator with voids for oxygen transport. With the immobilized electrolyte, VRLA batteries can reduce evaporation, leakage and vibration problems so that they can be operated in harsh conditions. It is noted that overcharging of lead–acid batteries generates oxygen and hydrogen. Flooded batteries have vents to release the gases and lose water while VRLA batteries recombine them to produce the water within the battery as:

$$2Pb + 2HSO_4^- + 2H^+ + O_2 \leftrightarrow 2PbSO_4 + 2H_2O \tag{1.24}$$

Hence, the water can be recycled in VRLA batteries during overcharge state. This feature leads to VRLA batteries achieving higher charging rate than flooded lead–acid batteries.

1.4.3 NiCd and NiMH Batteries

Nickel-based rechargeable batteries were developed more than 100 years ago. They offer higher performances than VRLA batteries, such as longer cycle life, larger capacity, and faster charging capability. Nickel-based batteries have been widely used in portable devices, telecommunication systems, aircraft and space satellite power systems, military applications, heavy industry and EVs, particularly in hybrid electric vehicles (HEVs). One drawback of nickel-based batteries is that they have relatively high self-discharge rates. The most common nickel-based batteries are NiCd and NiMH batteries .

1.4.3.1 NiCd Batteries

There are different types of NiCd batteries available commercially, such as pocket-plate, sintered-plate and fiber-plate types . The pocket-plate type is the most mature NiCd battery that provides very reliable performance. However, this type is heavy which limits its applications. The sintered-plate type is constructed in a thinner form to reduce battery internal resistance, achieving higher charging and discharging rates. Nevertheless, this type has the drawbacks of manufacturing complexity and high cost. The fiber-plate type uses polymer materials and a plating technique to eliminate most of the drawbacks of its predecessors.

NiCd batteries use nickel hydroxide ($Ni(OH)_2$) as a positive electrode, cadmium (Cd) as a negative electrode and potassium hydroxide solution (KOH) as an electrolyte. The overall chemical reaction is

$$2NiOOH + 2H_2O + Cd \leftrightarrow 2Ni(OH)_2 + Cd(OH)_2 \tag{1.25}$$

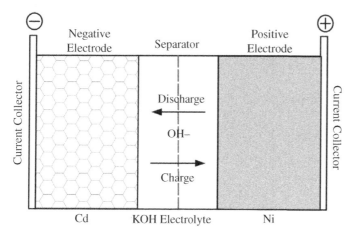

Figure 1.15 Schematic diagram of a NiCd battery.

It can be seen from Eq. (1.25) that metallic cadmium is oxidized to produce $Cd(OH)_2$ while NiOOH is reduced to produce $Ni(OH)_2$ during the discharge process. Figure 1.15 shows a schematic diagram of a NiCd battery. Unlike the sulfuric acid in lead–acid batteries, the concentration of the potassium hydroxide electrolyte in NiCd batteries only changes slightly during the charging and discharging processes and thus cannot be used to determine the SOC. One problem of NiCd batteries is the use of cadmium which has adverse effects on human beings and the environment.

1.4.3.2 NiMH Batteries

NiMH batteries have similar working principles to NiCd batteries. They use a hydrogen-absorbing alloy for a negative electrode instead of cadmium. Without using any cadmium, NiMH batteries are safer and more environmentally friendly. Furthermore, NiMH batteries provide higher energy density and longer cycle life than NiCd batteries. Gradually, NiMH batteries have replaced NiCd batteries in battery markets.

Figure 1.16 shows a schematic diagram of a NiMH cell. The positive electrode contains nickel hydroxide as its principal active material and the negative electrode is mainly composed of hydrogen-absorbing nickel alloys. The cell has an electrically insulating separator, an alkaline electrolyte (e.g. a solution of potassium hydroxide, KOH), and a vented metal case. There are many choices of metal hydride available. The most common ones are AB_2 alloy, a metal alloy of vanadium-titanium-zirconium-nickel and AB_5 alloy, a rare-earth metal alloy of lanthanum-nickel. The electrolyte does not enter into the anode reaction so that conductivity stays at a high level throughout the usable capacity of the battery. In addition, the nickel active material is insoluble in the electrolyte (KOH) which leads to longer life and better abuse tolerance. Only a proton is involved in the charge/discharge reaction, leading to very small changes of electrolyte density and improved mechanical stability of the anode during cycling. The overall chemical reaction for NiMH batteries is

$$MH + NiOOH \leftrightarrow M + Ni(OH)_2 \tag{1.26}$$

Inside NiMH batteries, there is recombination mechanism that allows oxygen to react with hydrogen at the negative electrode, producing water and minimizing the pressure

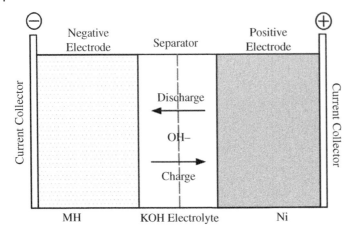

Figure 1.16 Schematic diagram of a NiMH battery.

of the released gases. Note that the charging rate should be controlled to ensure that the generated oxygen does not exceed the recombination rate.

1.4.4 Lithium-Ion Batteries

In order to achieve significant increase in improved energy density and power density, Li-ion batteries have been developed recently. Compared with lead–acid and nickel-based batteries, Li-ion batteries have higher charging and discharging capability, a longer cycle life, and no memory effect. They are very attractive for applications where weight or volume is important, such as mobile phones, laptop computers, and EVs. High initial cost has limited their use in price-sensitive applications, but new chemistries and economies of scale are promising to reduce the cost of Li-ion batteries. For example, large scale Li-ion battery systems have been gradually applied in stationary energy storage in power systems with high penetration of RESs [5].

Li-ion batteries use lithium intercalation compounds as the positive and negative electrodes, where lithium ions move between two electrodes. The negative electrode is deposited onto the copper foil, while the positive electrode, which is a mixture of electroactive material, carbon black and binder, is coated onto the aluminum foil. Both metallic foils ensure the function of the current collector. During discharging, lithium ions are released from the anode (negative electrode) and travel through an organic electrolyte toward the cathode (positive electrode). Electrons flow from the anode to the cathode though the external load. When lithium ions arrive at the cathode, they are quickly inserted into the cathode material. During charging, lithium ions are removed from the cathode for insertion into the anode material. The movement of lithium ions is also known as the intercalation process. Figure 1.17 shows a schematic diagram of a Li-ion battery. Li-ion batteries use materials with a layered structure and a three-dimensional framework structure as the positive electrode. The former includes lithium cobalt oxide (LCO), $LiCoO_2$, and lithium nickel manganese cobalt oxide (NMC), $LiNiMnCoO_2$, while the latter includes the lithium manganese oxide (LMO), $LiMn_2O_4$, and lithium iron phosphate (LFP), $LiFe_2PO_4$. In general, the materials at the negative electrode (anode) are various forms of carbon, particularly graphite and

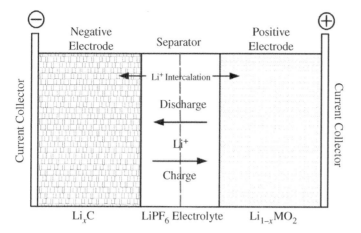

Figure 1.17 Schematic diagram of a Li-ion battery.

hydrogen-containing carbon materials. Different types of carbon electrodes in Li-ion batteries lead to various lithium intercalation capacities and performances. Liquid, gel, polymer or ceramic electrolyte has been adopted in Li-ion batteries. In general, the performance of cobalt oxide is the best among the three types while its cost is much higher than that of nickel oxide and manganese oxide.

There is an exceptional case where the negative electrode uses lithium titanate ($Li_4Ti_5O_{12}$) on the surface of its anode instead of carbon and this type of the battery is known as a lithium titanate battery. Since the lithium titanate battery makes the surface of the negative electrode significantly larger than that with the carbon, it allows electrons to interchange quickly and hence results in a faster charging rate than other types of Li-ion batteries. Its extreme fast charging characteristics will be particularly discussed in Chapter 6.

Although Li-ion batteries consist of different chemical reactions upon various combinations of adopted materials, their overall chemical reaction can be represented by

$$LiYO_2 + C \leftrightarrow Li_xC + Li_{1-x}YO_2 \tag{1.27}$$

where Y represents cobalt, nickel, or manganese. Take Y as cobalt for example, then x depends on the morphology of the cobalt oxide. Despite many advantages, Li-ion batteries are not as safe as lead–acid and nickel-based batteries. Several safety devices with shot-down separator for over temperature, tear-away tab for internal pressure, venting for pressure relief and thermal interruption for overcurrent have to be built into Li-ion batteries.

1.4.5 Battery Performance Comparison

Lead–acid, NiCd, NiMH, and Li-ion batteries have been used as energy sources to power EVs. Their key characteristics in terms of nominal voltage, specific energy and energy density, capacity efficiency and energy efficiency, specific power and power density, self-discharge, cycle life, and temperature operation range are discussed in the following [2, 16].

1.4.5.1 Nominal Voltage

The theoretical voltage of batteries is dependent on the electrode materials used in the batteries. Such a theoretical value may not be able to be achieved in real batteries. Furthermore, the practical voltage of batteries is varied during the charging and discharging processes. The notion of nominal voltage is adopted to define the terminal voltage of rechargeable batteries. The nominal voltage refers to the average value of the working voltage at constant current (CC) during the whole discharging process from a fully charged state to a fully discharged state.

Based on the theoretical voltage of each type of battery, the nominal voltage of a lead–acid cell and a nickel-based cell is defined as 2 and 1.2 V, respectively, and the nominal voltages of Li-ion battery cells are varied from 3.2 to 3.8 V, whilst a LCO has a nominal voltage of 3.6 V, a LMO cell 3.7 (or 3.8) V, a lithium phosphate cell 3.2 V and a lithium titanate cell 2.4 V.

1.4.5.2 Specific Energy and Energy Density

The specific energy is the ratio of the total amount of energy (Wh) stored in the active materials inside a battery cell to the mass of the battery cell (kg). The energy density is the ratio of the total amount of energy (Wh) stored in the active materials inside a battery cell to the volume of the battery cell (l).

Li-ion batteries have the highest specific energy of 250 Wh kg^{-1} which is about 3, 2.6 and 6 times higher than that of NiCd, NiMH and lead–acid batteries, respectively. They also have the highest energy density of 693 Wh l^{-1} which is about 4.6, 2.3 and 6.3 times higher than that of NiCd, NiMH and lead–acid batteries, respectively.

1.4.5.3 Capacity Efficiency and Energy Efficiency

The ability of batteries to accept charge or be recharged is measured in terms of the coulombic efficiency and the energy efficiency of the charge/discharge cycle. The coulombic efficiency is defined as

$$\mu_c = \int_0^{t_{dis}} i_{dis} dt / \int_0^{t_{cha}} i_{cha} dt \tag{1.28}$$

where t_{dis} and t_{cha} are the total discharge and charge times, respectively, and i_{dis} and i_{cha} are the battery currents flowing during the discharging and charging processes, respectively. The energy efficiency is defined as

$$\mu_e = \int_0^{t_{dis}} v_{dis} i_{dis} dt / \int_0^{t_{cha}} v_{cha} i_{cha} dt \tag{1.29}$$

where v_{dis} and v_{cha} are the battery voltages during the discharging and charging processes, respectively.

Since the voltage in the charging process is higher than that in the discharging process, energy efficiency is lower than capacity efficiency. For example, the energy efficiency of Li-ion batteries is generally around 85–95% and their capacity efficiency is near to 100%.

1.4.5.4 Specific Power and Power Density

The specific power is the ratio of the rated power (W) which can be delivered from a battery cell to the mass of the battery cell (kg). The power density is the ratio of the rated power (W) which can be delivered from a battery cell to the volume of the battery cell (l). They both indicate how quickly the batteries can provide the necessary energy to loads or how much energy the batteries can deliver in a particular time. Li-ion batteries have the highest specific power and power density, followed by NiMH and then NiCd batteries; lead–acid batteries are the least powerful.

1.4.5.5   Self-discharge

Self-discharge is caused by internal chemical reactions. It reduces the stored charge of batteries even when batteries are not in use. All batteries have internal leakage (self-discharge) and the leakage increases with temperature. Li-ion batteries and lead–acid batteries both have a self-discharge rate of 5% per month compared with 10% per month for NiCd batteries and 20% per month for NiMH batteries.

1.4.5.6   Cycle Life

The cycle life is the number of charge/discharge cycles that can be achieved before a battery reaches the end of its useful life. In EVs, the end of useful life is defined as a capacity drop to 80% of the original capacity. Cycle life depends mainly on battery temperature, discharge profile and depth of discharge (DOD). In general, batteries have longer life for low DOD cycles. For example, Li-ion batteries typically last 2000 cycles at low discharge/charge rates and room temperature at 100% DOD; they can last up to 20 000 cycles at 20–40% DOD.

1.4.5.7   Temperature Operation Range

Batteries perform poorly at extremely low and high temperatures. Low temperatures hinder ionic diffusion and migration and lead to damaging side reactions (e.g. lithium plating). High temperatures cause other side reactions, such as corrosion and gas generation. For lead–acid batteries, charge and discharge temperatures should be limited to an operation range of −15 to 50 °C. Li-ion batteries have an operation range of −20 to 60 °C. NiCd and NiMH batteries have operation ranges of −20 to 50 °C and −20 to 60 °C, respectively.

   Table 1.4 shows the comparison of lead–acid, NiCd, NiMH and Li-ion batteries. It can be seen that Li-ion batteries have advantages over the other three types of batteries in almost all aspects. Furthermore, most EVs today use Li-ion batteries [14, 15]. Thus, the focus will be on advanced BMSs for Li-ion batteries for the remainder of the book.

1.5   Key Battery Management Technologies

The development of advanced BMSs for EVs involves many technologies due to the complicated electrochemical reactions and diversity of the cells in the battery system as well as their performance degradation over time. In the following, the key BMTs will be discussed. They include battery modeling, battery states estimation, battery charging, and battery balancing [2, 8, 9].

1.5.1   Battery Modeling

Battery modeling forms the foundation for the design, control and optimization of BMSs in EVs. It has a significant impact on the estimation of battery states, battery balancing,

Table 1.4 Comparison of four types of batteries in EVs.

	Lead–acid	NiCd	NiMH	Li-ion
Specific energy (Wh kg^{-1})	33–42	50–80	70–95	118–250
Energy density (Wh l^{-1})	60–110	50–150	140–300	250–693
Specific power (W kg^{-1})	180	200	200–300	200–430
Power density (W l^{-1})	450	200	300	800
Nominal voltage (V)	2	1.20	1.20	3.6
Overcharge tolerance	High	Moderate	Low	Very low
Self-discharge (per month)	<5%	10%	20%	<5%
Operating temperature (°C)	−15 to 50	−20 to 50	−20 to 60	−20 to 60
Cycle life	500 ∼ 1000	2000	<3000	2000
Energy efficiency	>80%	75%	70%	85–95%

battery charging, and battery fault diagnosis. Over the years, numerous battery models have been developed to describe battery behaviors. All these models can be mainly categorized into three groups: electrochemical models; blackbox models; and equivalent circuit models.

Electrochemical models provide a highly accurate description of electrochemical processes inside a battery. The models involve many partial differential equations (PDEs). Solving these PDEs leads to large computation burden overhead. Furthermore, many battery electrochemistry related parameters in PDEs are practically difficult to acquire in EVs. In general, electrochemical models are suitable for battery design and simulation but are not appropriate for real-time BMSs in EVs.

Blackbox models apply artificial intelligence (AI) techniques to capture the relationship between the selected inputs and outputs of batteries without the prior knowledge of electrochemical processes inside batteries. AI techniques such as neural network (NN) [19], fuzzy logic (FL) [20], and support vector machine (SVM) [21] apply training approaches to establish the models based on testing data. To achieve high model accuracy and good generalization ability, testing data should cover sufficient battery operation ranges, and the parameters in the training approaches need to be effectively tuned.

Equivalent circuit models describe battery behaviors using a combination of the lumped circuit parameters, such as resistances, capacitances and voltage or current sources. Due to their simple model structure and relatively small number of model parameters, the models are widely adopted in real-time BMSs for EVs, where overall dynamic characteristics of battery terminal voltage, current, temperature, and SOC are of more interest than the detailed electrochemical reactions inside batteries. Furthermore, equivalent circuit models are more intuitive and can be easily incorporated in the electrical circuits in BMSs. Thus, the focus of this book is on the equivalent circuit models. Chapter 2 will provide detailed discussions of battery modeling techniques.

1.5.2 Battery States Estimation

Battery states are crucial for BMSs to monitor and protect batteries and optimize the operation of batteries. Many battery parameters can indicate battery states. The discussions will be on the key states used in BMSs, including SOC, state of energy (SOE), state of power (SOP), and state of health (SOH).

The SOC is the ratio of the remaining capacity to the nominal capacity of the battery. Similarly, the SOE is the ratio of the remaining energy to the nominal energy of the battery. They are both often expressed as a percentage. If the battery is fully charged, the SOC (or SOE) is known as 100%; if the battery is fully discharged, the SOC (or SOE) is defined as 0%. Numerous methods are developed for SOC (or SOE) estimation. Generally, they are categorized as direct estimation methods and model-based estimation methods. For the direct estimation methods, the SOC (SOE) can be calculated by Ampere-hour (Ah) counting and open circuit voltage (OCV) at battery equilibrium state. For the model-based estimation methods, they can be further divided into two groups. In the first group, electrochemical models and equivalent circuit models are usually represented in the form of state space equations, where the SOC (or SOE) is one of the state variables. Then, state observers are adopted for the SOC (or SOE) estimation, such as Kalman filters (KFs) and their variants, sliding mode observers, and H infinite filters. In the second group, blackbox models are used to directly establish the relationship between the measurable battery parameters (i.e. voltage, current, and temperature) and the SOC (or SOE) based on testing data through training approaches, such as NN, FL and SVM models. Chapter 3 is devoted to presenting the SOC and SOE estimation approaches in detail.

The SOH is the ratio of the capacity of the aged battery to the nominal capacity of the new battery. It is also expressed as a percentage. For the new battery, the SOH is defined as 100%. When the SOH falls to 80%, it is defined as the end of the battery life for EVs. However, EV batteries at a SOH of 80% may be able to be reused as stationary energy storage. Numerous methods have been proposed to estimate the SOH; they are categorized as the model-free estimation method, the model-based estimation method, and the data-driven estimation method. For the model-free estimation method, the standard capacity test is conducted to measure the aged battery capacity in each charge and discharge cycle. According to the definition of the SOH, it can be calculated by dividing the measured capacity by the nominal capacity. For the model-based estimation method, they can be further divided into two groups. In the first group, battery capacity is taken as a slow time-varying parameter based on equivalent circuit models [22] and electrochemical models [23]. Various observer techniques such as KFs and their variants and sliding mode observers are adopted to estimate the capacity for the calculation of the SOH. In the second group, battery life-cycle and aging models are developed to estimate the capacity degradation and predict the battery life cycle [24]. For the data-driven estimation method, AI techniques are employed to analyze the measured battery parameters such as battery terminal voltage, current, and temperature in real time for the SOH estimation; these AI techniques include SVM, NN, Naive Bayes, Bayesian learning and Bayesian Monte Carlo methods [25]. There is another way to calculate the SOH based on the ratio of the internal resistance of the aged battery to the internal resistance of the new battery. The above-mentioned model-free, model-based and data-driven methods can also be used to estimate the SOH in terms of battery internal resistance. The details of the SOH estimation approaches are discussed in Chapter 4.

The SOP is defined as the ratio of peak power to nominal power. The accurate estimation of peak power in BMSs is critical in EVs since it is necessary to determine power available at the moment to meet acceleration, regenerative braking and gradient climbing power requirements without overcharging or over-discharging batteries [26]. More importantly, the SOP can be used to optimize the relationship between battery capacity and EV performances. Many SOP methods are developed for EVs. They can be divided into four types. The first type is the hybrid pulse power characterization (HPPC) method. When the HPPC method is applied to estimate the SOP, it only considers operation voltage limits while estimating peak power and instantaneous current. The second type is the SOC-limited method. It takes into account the maximum and minimum operation SOCs of a battery. Generally, it provides an optimistic prediction of peak current to be discharged or charged over a wide range of the SOC. The third type is the voltage-limited method. It is suitable for continuous peak power estimation while neglecting the SOC limits. The fourth type is the multi-constraint dynamic method. This method estimates peak power under the constraints of its voltage, current, temperature, available capacity, and SOC. The relaxation effect is also considered in this method. Chapter 5 is devoted to discussing the details of SOP estimation approaches.

1.5.3 Battery Charging

Battery charging plays an important role in the wide adoption of EVs. It will refill the battery system on its depletion. The issues on battery charging involve charging time, power transfer and charging facilities, and charging algorithm. Among them, the charging algorithm is crucial to battery charging. It defines a charging profile to charge the battery system in terms of current and voltage over time and a condition to terminate the charging process. The charging profile of CC is normally used to charge NiCd/NiMH batteries, whereas the charging profile of constant voltage (CV) is normally used to charge lead–acid batteries. For Li-ion batteries, many charging algorithms have been developed, including constant current constant voltage (CC/CV), multistep constant current (MSCC), two-step constant current constant voltage (TSCC/CV), constant voltage constant current constant voltage (CVCC/CV), pulse current (PC), and pulse voltage (PV) [18]. Recently, there have been two new developments in charging Li-ion batteries. One was to find the optimal charging current profile for Li-ion batteries and the other was to develop a Li-ion battery with extremely fast charging and discharging capability.

For the termination of a charging process, SOC is an ideal indicator to stop charging a battery when the SOC reaches 100%. However, the SOC is not a directly measured parameter. The accurate estimation of the SOC is a very challenging task. Even if the SOC can be estimated accurately, backup methods are still required to stop a charging process using the directly measured parameters, such as voltage, current, or the temperature of a battery. Different techniques have been proposed to stop the charging process in terms of battery electrochemical characteristics. The voltage drop (ΔV) is normally incorporated with the CC to stop charging NiCd/NiMH batteries, whereas the cut-off current is normally incorporated with the CV to stop charging lead–acid and Li-ion batteries. There are many other termination techniques, such as timer, temperature cut-off, temperature change rate (dT/dt), cut-off voltage for charge, cut-off voltage for discharge and voltage change rate (dV/dt). Chapter 6 discusses battery charging techniques in detail.

1.5.4 Battery Balancing

Battery systems generally consist of hundreds and thousands of battery cells connected in series and parallel to fulfill voltage, power and energy requirements for EVs. An imbalance of the serially connected cells in battery systems occurs inevitably after a few charge and discharge cycles and deteriorates over time. The imbalanced conditions can cause overcharging or over-discharging cells, prevent the full utilization of battery capacity and reduce lifecycle of battery systems. Battery balancing is required to equalize voltage, SOC and capacity of all cells in battery systems to maintain their performances [27].

Battery system imbalance can be capacity imbalance, SOC imbalance, and voltage imbalance. Capacity imbalance takes place in the aged battery system. It mainly results from different aging speeds of the cells which are strongly affected by the charging and discharge current, DOD, and ambient temperature. SOC imbalance takes place in the fresh battery system. It is caused by different charging and discharge efficiencies and self-discharge rates of the cells. Voltage imbalance takes place in the battery system at any time. It is caused by cell capacity divergences and different internal parameters such as internal resistance.

There are two categories of battery balancing methods. One category is the passive balancing method. It equalizes the battery system by dissipating the energy from cells into heat through the shunting resistor and is only effective in the charging process to help fully charge all the cells in the battery system. The other category is the active balancing method. It equalizes the battery system by transferring the energy among cells in the battery system through power electronic circuits and can be used in both charging and discharging processes to help fully charge and discharge all the cells in the battery system. Chapter 7 gives a detailed discussion of battery balancing techniques.

1.6 Battery Management Systems

The capacity and voltage of a single battery cell is relatively low. A battery system is required in EVs. The battery system consists of hundreds and thousands of single cells connected in series and parallel, where the series connection of cells yields a higher total battery voltage at the same capacity and the parallel connection of cells yields a higher total battery capacity at the same battery voltage. To manage such a large number of battery cells, a BMS is highly desirable, which should have an ability to maintain the health of all the cells to deliver the power and energy as required by EVs through monitoring and controlling the cells, estimating battery states, balancing the cells, and reporting the status of the cells in battery systems [7–9].

The BMS in EVs consists of many sensors, actuators, and controllers embedded with models and algorithms. The main tasks of the BMS are as follows:

- Monitor and measure the parameters of all the cells in battery systems.
- Protect the battery system to ensure its safe and reliable operation.
- Maintain all the cells within the manufacturer-recommended operating conditions to prolong the life of the battery system.
- Optimize the energy usage of the battery system.
- Communicate to other units and the vehicle control unit (VCU).

These tasks are implemented by the hardware and software built in the BMS.

1.6.1 Hardware of BMS

The major hardware circuits of the BMS include: (i) current, voltage and temperature sensors to measure battery parameters. (ii) Battery state estimators to estimate battery states, such as SOC and SOH. (iii) On-board fault diagnosis to detect the faults including sensor fault, actuator fault, communication network fault, battery fault, loose connection, exceeding combustible gas concentration, insulation fault, uniformity fault, over-fast temperature rise fault, and provide the warning signals of these faults. (iv) Battery protection circuits to prevent the battery system from overcharging, over-discharging, overcurrent, ultra-high temperature, and ultra-low temperature. When the corresponding threshold values are exceeded the error messages are provided to alert EV drivers and correspondingly the safety switches execute protection commands. (v) Battery passive or active equalization circuits (equalizers) to balance the voltages or SOCs or capacities among the cells as consistent as possible. (vi) Battery charger to charge the battery system based on the predefined charging algorithm. (vii) Thermal controller to start the heater or fan during battery charging or discharging processes. (viii) Communication unit to transfer measured data and conduct on-line calibrating, monitoring, automatic code generation, and program downloading. (ix) Information storage unit to store key data, such as SOC, SOH, internal resistance, maximum and minimum voltage and temperature, accumulated charge and discharge ampere hours, fault code and uniformity.

1.6.2 Software of BMS

The software of the BMS is considered as a "brain" which will control the operation of the hardware, make decisions, and estimates states. It includes: (i) battery models to represent the cell and battery system. (ii) Estimation algorithms for battery states, such as SOC (SOE), SOH, and SOP. (iii) Fault diagnosis algorithms to detect different types of faults. (iv) Control algorithms to control the thermal system in terms of the normal range of working temperatures and the safety switches in terms of the range of normal working voltages. (v) Charging algorithms to decide the level of voltage and current to charge the battery systems. (vi) Battery balancing algorithms to equalize the battery cells based on battery terminal voltage or SOC or capacity as a balancing criterion. (vii) Communication protocols to control the data flows within the BMS and to exchange the data with VCU and other parts of EVs including a power inverter/charger, environmental control units (e.g. fans and heaters). Figure 1.18 illustrates the functional blocks of BMSs for EVs.

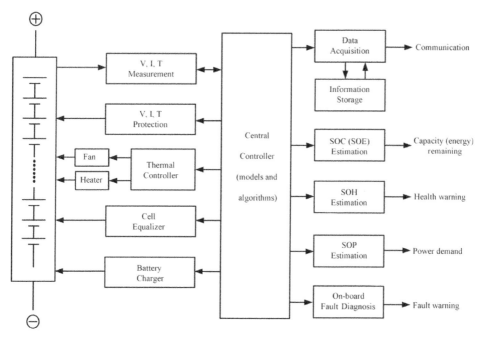

Figure 1.18 Battery management system for EVs.

To explain how these blocks work together, the process of battery charging in the BMS is taken as an example. During the charging process, the measurement block converts the currents, voltages, and temperatures of each cell in the battery system into digital signals. These measured parameters are sent to the state estimation block to estimate SOC and SOH and to the thermal system to heat up or cool the battery systems. The estimated SOC and SOH are sent to the battery cell equalizer and charge controller for the calculation of balancing and charging currents and charging voltage based on the battery model. The equalizer and charge controller generate the control signals of the balancing currents, charging currents, and charging voltage for the charger to perform cell balancing while charging the battery systems according to a pre-selected charging algorithm such as the constant current constant voltage charging algorithm. Once the battery system is fully charged in terms of the preset charging termination parameters, such as 100% SOC or the upper cut-off battery, the safety switches are turned off to protect the battery system from overcharging. The communication block is responsible for transferring the data related to the charging algorithm inside the BMS.

There are a number of different BMS architectures depending on the number of cells in the battery system and its physical structure. These architectures are mainly classified into two categories [28]: the centralized BMS and the distributed BMS.

1.6.3 Centralized BMS

The centralized BMS is a single controller board to monitor and control all the cells in the battery system. It requires extensive wiring from all the cells to the controller board and thus can hardly be scaled up. The centralized BMS is the most economical architecture and is suitable for a battery system with a small number of cells.

1.6.4 Distributed BMS

The distributed BMS has a combination of many slave boards and a master board. A slave board measures and controls an individual cell in the battery system. The slave boards are interconnected via a serial network and controlled by the master control board. A distributed BMS architecture is much easier to install than a centralized one. For a battery system with a large number of cells, the cost of a distributed system can be significantly higher than for a centralized system.

Chapter 8 will describe the structure of the BMS, its representative products and the key technologies in the future generation of the BMS.

1.7 Summary

This chapter describes EV fundamentals including the concepts of traction force, aerodynamic drag force, rolling resistance force and climbing force applied to EVs, focusing on the derivation of governing equations of EV motion. Based on these equations, the power and energy demands of EV batteries are discussed in terms of EV driving range per charge under standard driving cycles, acceleration rate, and maximum speed.

Then, an introduction to electrochemistry of batteries is provided to lay the foundation to explain the working principles of the most common batteries in EVs: lead–acid, NiCd, NiMH, and Li-ion batteries. The performances of these batteries are compared and discussed.

Finally, the key BMTs including battery modeling techniques, battery SOC, SOE, SOH and SOP estimation approaches and their implementation in BMS are briefly presented. The compositions, functions and architectures of the BMS are also discussed.

References

1 Chan, C.C. (2007). The state of the art of electric, hybrid, and fuel cell vehicles. *Proceedings of the IEEE* 95 (4): 704–718.

2 Kumar, M.S. and Revankar, S.T. (2017). Development scheme and key technology of an electric vehicle: an overview. *Renewable and Sustainable Energy Reviews* 70: 1266–1285.

3 Chrisafis, A. and Vaughan, A. (2017) France to ban sales of petrol and diesel cars by 2040. https://www.theguardian.com/business/2017/jul/06/france-ban-petrol-diesel-cars-2040-emmanuel-macron-volvo (accessed 24 April 2018).

4 *The Economist* (2017) China moves towards banning the internal combustion engine. https://www.economist.com/news/business/21728980-its-government-developing-plan-phase-out-vehicles-powered-fossil-fuels-china-moves (accessed 24 April 2018).

5 Garcia-Valle, R. and Pecas Lopes, J.A. (2013). *Electric Vehicle Integration into Modern Power Networks, Power Electronics and Power Systems*. New York: Springer.

6 Deng, J., Li, K., Laverty, D.M. et al. (2014). Li-ion battery management system for electric vehicles-A practical guide. In: *International Conference on Life System Modeling and Simulation and International Conference on Intelligent Computing for Sustainable Energy and Environment*, 32–44. Springer.

7 Rahimi-Eichi, H., Ojha, U., Baronti, F., and Chow, M.Y. (2013). Battery management system, an overview of its application in the smart grid and electric vehicles. *IEEE Industrial Electronics Magazine* 7 (2): 1–16.

8 Hannan, M.A., Lipu, M.S.H., Hussain, A., and Mohamed, A. (2017). A review of lithium-ion battery state of charge estimation and management system in electric vehicle applications: challenges and recommendations. *Renewable and Sustainable Energy Reviews* 78: 834–854.

9 Lu, L., Han, X., Li, J. et al. (2013). A review on the key issues for lithium-ion battery management in electric vehicles. *Journal of Power Sources* 226: 272–288.

10 Chau, C.C. and Chau, K.T. (2001). *Modern Electric Vehicle Technology*. Oxford: Oxford University Press.

11 Zhang, X. and Mi, C. (2001). *Vehicle Power Management, Modeling, Control and Optimization*. London: Springer.

12 Wikipedia (2018) Driving cycle. http://en.wikipedia.org/wiki/Driving_cycle (accessed 24 April 2018).

13 Giakoumis, E.G. (2017). *Driving and Engine Cycles*. Cham, Switzerland: Springer.

14 Berckmans, G., Messagie, M., Smekens, J. et al. (2017). Cost projection of state of the art lithium-ion batteries for electric vehicles up to 2030. *Energies* 10: 1314.

15 Grunditz, E.A. and Thiringer, T. (2016). Performance analysis of current BEVs-based on a comprehensive review of specifications. *IEEE Transactions on Transportation Electrification* 2: 270–289.

16 Rahn, C.D. and Wang, C.Y. (2013). *Battery Systems Engineering*. Chichester: Wiley.

17 Berndt, D. (1997). *Maintenance-Free Batteries-Lead-Acid, Nickel/Cadmium, Nickel/Metal Hydride a Handbook of Battery Technology*, 2e. Taunton: Research Studies Press Ltd.

18 Chau, K.T. (2016). *Energy Systems for Electric and Hybrid Vehicles*. London: The Institution of Engineering and Technology.

19 Shen, W.X., Chan, C.C., Lo, E.W.C., and Chau, K.T. (2002). A new battery available capacity indicator for electric vehicles using neural network. *Energy Conversion and Management* 43 (6): 817–826.

20 Chau, K.T., Wu, K.C., Chan, C.C., and Shen, W.X. (2003). A new battery capacity indicator for nickel–metal hydride battery powered electric vehicles using adaptive neuro-fuzzy inference system. *Energy Conversion and Management* 44 (13): 2059–2071.

21 Wang, J., Chen, Q., and Cao, B. (2006). Support vector machine based battery model for electric vehicles. *Energy Conversion and Management* 47 (7): 858–864.

22 Xiong, R., Sun, F., and He, H. (2016). Model-based health condition monitoring method for multi-cell series-connected battery pack. In: *IEEE Transportation Electrification Conference and Expo (ITEC)*, 1–5. IEEE.

23 Xiong, R., Li, L., Li, Z. et al. (2018). An electrochemical model based degradation state identification method of lithium-ion battery for all-climate electric vehicles application. *Applied Energy* 219: 264–275.

24 Zhang, Y., Xiong, R., He, H., and Pecht, M. (2018). Lithium-ion battery remaining useful life prediction with Box-Cox transformation and Monte Carlo simulation. *IEEE Transactions on Industrial Electronics* https://doi.org/10.1109/TIE.2018 .2808918.

25 Zhang, Y., Xiong, R., He, H., and Pecht, M. (2018). Long short-term memory recurrent neural network for remaining useful life prediction of lithium-ion batteries. *IEEE Transactions on Vehicular Technology* https://doi.org/10.1109/TVT.2018 .2805189.

26 Sun, F., Xiong, R., and He, H. (2014). Estimation of state-of-charge and state-of-power capability of lithium-ion battery considering varying health conditions. *Journal of Power Sources* 259: 166–176.

27 Cu, X., Shen, W.X., Zhang, Y. et al. (2016). Novel active LiFePO$_4$ battery balancing method based on chargeable and dischargeable capacity. *Computers and Chemical Engineering* 97: 27–35.

28 Bingeman, M. and Ben Jeppesen, N. B. (2016) Improving battery management system performance and cost with Altera FPGAs. WP-01247-1.1, Altera Corporation.

2

Battery Modeling

2.1 Background

Battery modeling plays an important role in estimating battery states which include state of charge (SOC), state of health (SOH), state of energy (SOE), and state of power (SOP). These states can hardly be measured directly and have to be estimated based on battery models. The batteries in electric vehicles (EVs) are operating in a complex and highly dynamic environment, which means they possess highly time-varying characteristics. Electrochemical models (EMs), black box models, and equivalent circuit models (ECMs) have been used to describe such characteristics. EMs apply partial differential equations (PDEs) to describe the electrochemical reaction process inside batteries [1]. These models generally have high accuracy but their computation burden is too heavy to integrate into on-board battery management systems (BMSs) for EVs. Black box models such as neural network (NN) models require a large amount of experimental data to train the models. Since the parameters of the trained NN models do not have physical meanings, the obtained NN models can only be used in similar working conditions that the experimental data are collected from [2]. This limits the applications of the NN models into on-board BMSs for EVs. ECMs use electrical circuits to represent battery terminal behaviors [3, 4]. They are compatible with the circuits of BMSs for EVs and can be easily embedded into BMSs. This chapter first provides a brief introduction of EMs and black box models, and then explains ECMs as well as the methods to identify the parameters of ECMs in detail.

2.2 Electrochemical Models

EMs describe physical and electrochemical processes in the anode, separator, and cathode regions of batteries by using thermodynamics and electrochemical kinetics equations [5]. Based on the theories of porous electrodes and concentrated solutions, a pseudo-two-dimensional (P2D) model has been proposed to model lithium-ion batteries (LiBs). It consists of coupled nonlinear PDEs, including the diffusion equation of lithium ions inside active particles and electrolytes, the solid phase balance equation, and the electrolyte phase balance equation in three regions of LiBs. The internal structure of a LiB used to derive the P2D model is shown in Figure 2.1. A LiB is composed of porous electrodes, the separator, and current collectors. The negative electrode is mainly made of carbon, graphite and silicon. The positive electrode is mainly made of

Advanced Battery Management Technologies for Electric Vehicles, First Edition. Rui Xiong and Weixiang Shen.
© 2019 John Wiley & Sons Ltd. Published 2019 by John Wiley & Sons Ltd.

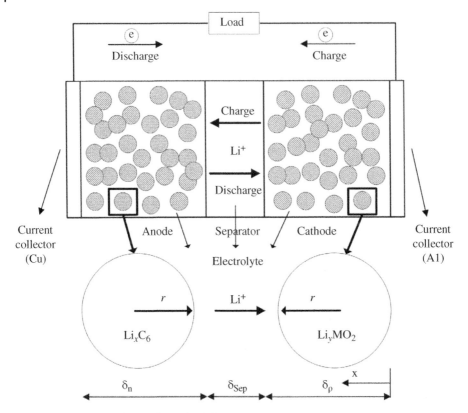

Figure 2.1 Internal structure of a LiB for P2D.

metal oxide. The separator has a porous structure. In the first charge/discharge cycle, an ion-conducting protective but electronically insulating passivation layer, also known as the solid electrolyte interphase (SEI), is formed on the surface of the anode due to the interface reaction between the anode and the electrolyte.

Lithium ions diffuse in the electrolyte through the SEI while electrons are transported in the external circuit. The P2D model contains two dimensions. One dimension is from the negative electrode toward the positive electrode along the x-axis and the other is the radial dimension of the spherical particle along the r-axis. The particles are considered to be distributed throughout the electrodes and are embedded at each position along the x-axis. In the P2D model, the adjacent particles are not directly coupled.

During the discharge process, an internal electrochemical reaction process can be described as follows. At the anode, active particles react with the electrolyte at the electrolyte interface and the concentration of lithium ions decreases at the surface of the active particles, resulting in a concentration gradient inside the active particles and the diffusion of lithium ions from inside of the particles to the surface of the particles. Subsequently, lithium ions deintercalate from the interface of the active particles, releasing lithium ions into the electrolyte. These lithium ions diffuse through the separator to the cathode. At the cathode, lithium ions are embedded in the surface of the active particles, leading to the formation of a lithium ion concentration gradient inside the active particles. Then, the lithium ions diffuse and migrate from the outside to the inside of the active particles. Simultaneously, the electrons compensate the charge balance. Since the porous separator is an electron insulator, electrons have to flow in the external circuit through a load, which converts chemical energy into electrical energy. During the charge process, lithium ions and electrons travel in the reverse direction.

According to the P2D model, the dynamic characteristics of LIBs can be simplified into the following three parts: first, lithium ions can be described by the equations for the conservation law of matter and the conservation law of electrons in the internal solid phase and liquid phase diffusion process. Secondly, the potential of the solid phase in the positive and negative electrodes and the liquid phase in the electrolyte can be described by the conservation of the charge equation. Thirdly, the electrochemical reaction that occurs on the surface of the active particles can be described by Butler–Volmer (BV) kinetics. Although the P2D model can fully describe the physical-electrochemical reaction inside a battery, the model contains a large number of PDEs and model parameters. To improve the model efficiency, it is necessary to reduce order and simplify the P2D model. The one-dimensional model and single particle model are two examples of the reduced P2D model. Although EMs can accurately describe dynamic behaviors and battery states, the complexity of the model and the excessive consumption of resources hinder their application in BMSs for EVs. However, they are still used to develop charging strategies, thermal management, and energy management of batteries through simulation or in the laboratory environment with high computing power.

2.3 Black Box Models

Black box models can simulate a complex relationship between external parameters of batteries without knowing their internal electrochemical reaction process, such as the relationship of the SOC to battery terminal voltage and discharge current. Although batteries appear deceptively simple, they are very sophisticated electrochemical systems as elaborated in EMs. First, they embody a set of interacting physical and chemical processes which convert chemical energy into electrical energy and vice versa. Secondly, besides the main electrochemical reaction, there are also some side reactions, such as corrosion and self-discharge. Thirdly, battery performance is greatly influenced by environmental conditions as well as charging and discharging current profiles. The black box models, such as NN models, mimic the learning process of a human brain to obtain the key patterns within a multi-dimensional information domain through automatic training. A NN model consists of an input layer, some hidden layers, and an output layer. Inside the NN, each single neuron is connected to other neurons of a previous layer through adaptable weights. Knowledge is usually stored as a set of connection weights. Training is the process of modifying the connection weights in some orderly fashion using a learning algorithm, in which the input is presented to the NN along with the desired output, and the weights are then adjusted so that the NN attempts to produce the desired output. Typically, the learning process training is stopped when the average error between the desired and actual outputs of the NN model over the training data sets is less than a predetermined threshold. The learning time is dictated by various factors, including the complexity of the problem, the number of training data, the structure of

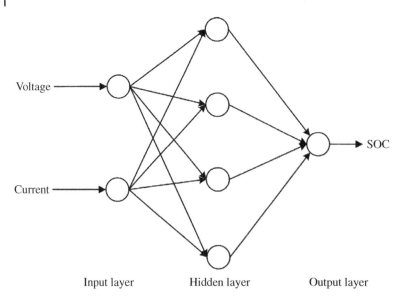

Figure 2.2 NN model for the relationship of SOC to terminal voltage and current for a battery.

the NN model, the learning parameters, and the learning algorithm. Figure 2.2 shows a three-layer NN model to describe the relationship between the SOC and terminal voltage as well as current [6]. In this NN model, terminal voltage and current are in the input layer, four neurons in the hidden layer, and the SOC in the output layer. Battery temperature, current, and internal resistance can also be used as the inputs while battery available capacity can be chosen as an output [7].

During the modeling process, since the NN model for battery capacity is independent of battery type, it can readily be applied to any other types of batteries provided that the corresponding experimental data are available to train the NN model. In addition to the NN models, support vector machines [8] and fuzzy learning [9] can also be used to build black box models. A large amount of experimental data is required for training to obtain accurate black box models. Under the influence of uncertain operating conditions such as temperature and driving cycles of EVs, it is difficult to obtain a sufficient experimental data in advance to train black box models for BMSs in real EVs [10].

2.4 Equivalent Circuit Models

ECMs have been studied for BMS development in EVs. They are the lumped models with relatively few numbers of parameters. The ECM can depict static and dynamic characteristics. The static characteristic is described by equilibrium potential or open circuit voltage (OCV), which is represented by an ideal voltage source. The OCV is usually monotonically related to the SOC, aging levels, and operating temperatures; it is usually used to estimate SOC. The dynamic characteristic is described by the combination of polarization and hysteresis voltages with ohmic resistance and resistance–capacitor (RC) networks.

2.4.1 General *n*-RC Model

Based on the analysis of battery terminal voltage behavior, a general *n*-RC model is presented in Figure 2.3 [11]. R_i represents the internal resistance which is the sum of electrode material resistance, electrolyte resistance, diaphragm resistance, and contact resistance. RC networks denote the polarization effects with polarization resistance R_{Di} and polarization capacitor C_{Di}, $i = 0, 1, \ldots, n$. i_L is the load current (positive for discharge and negative for charge), U_t is the terminal voltage, and U_{Di} is the polarization voltage.

According to Kirchhoff's current law, the relationship between output voltage and input current can be expressed by

$$U_t(s) = U_{oc}(s) - i_L(s)\left(R_i + \frac{R_{D1}}{1 + R_{D1}C_{D1}s} + \cdots + \frac{R_{Dn}}{1 + R_{Dn}C_{Dn}s}\right)$$
$$(n = 0, 1, 2, \cdots\cdots) \tag{2.1}$$

which leads to the transfer function as

$$G(s) = \frac{U_t(s) - U_{oc}(s)}{i_L(s)} = -\left(R_i + \frac{R_{D1}}{1 + R_{D1}C_{D1}s} + \cdots + \frac{R_{Dn}}{1 + R_{Dn}C_{Dn}s}\right) \tag{2.2}$$

2.4.2 Models with Different Numbers of RC Networks

To apply the *n*-RC model into BMSs in EVs, the model shown in Eq. (2.2) is transferred from the frequency domain into the time domain. In the next section, the mathematical expressions of the models with different numbers of RC networks will be analyzed in the time domain.

2.4.2.1 R_{int} Model

When $n = 0$, the *n*-RC model is simplified to a R_{int} model, the relationship between the output voltage and the input current under this model is expressed as

$$G(s) = \frac{U_t(s) - U_{oc}(s)}{i_L(s)} = -R_i \tag{2.3}$$

$$U_{t,k} = U_{oc,k} - R_i i_{L,k} \tag{2.4}$$

where $U_{t,k}$ denotes the terminal voltage, $U_{oc,k}$ denotes the OCV, and $i_{L,k}$ stands for the load current at *k*th sampling time. We can reformulate Eq. (2.4) as

$$y_k = \Phi_{0,k}\theta_{0,k} \tag{2.5}$$

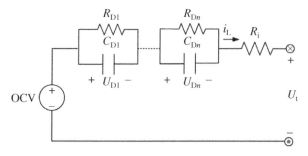

Figure 2.3 *n*-RC model for a LiB.

where $y_k = U_{t,k}$ is the system output variable. For the R_{int} model, $\mathbf{\Phi}_{0,k} = [1 \; i_{L,k}]$ is the system data matrix, and $\mathbf{\theta}_{0,k} = [U_{oc,k} - R_i]^{\mathrm{T}}$ is the system parameter vector.

2.4.2.2 Thevenin Model

When $n = 1$, the n-RC model is simplified to a Thevenin model

$$G(s) = \frac{U_t(s) - U_{oc}(s)}{i_L(s)} = -\left(R_i + \frac{R_{D1}}{1 + R_{D1}C_{D1}s}\right) \tag{2.6}$$

By defining $E_L(s) = U_t(s) - U_{oc}(s)$, we obtain

$$G(s) = \frac{E_L(s)}{i_L(s)} = -\frac{R_i + R_{D1} + R_i R_{D1} C_{D1} s}{1 + R_{D1} C_{D1} s} \tag{2.7}$$

A bilinear transformation method

$$s = \frac{2}{\Delta t} \frac{1 - Z^{-1}}{1 + Z^{-1}} \tag{2.8}$$

is employed for the discretization of Eq. (2.7), and we obtain

$$G(Z^{-1}) = -\frac{\dfrac{R_i \Delta t + R_{D1} \Delta t + 2R_i R_{D1} C_{D1}}{\Delta t + 2R_{D1} C_{D1}} + \dfrac{R_i \Delta t + R_{D1} \Delta t - 2R_i R_{D1} C_{D1}}{\Delta t + 2R_{D1} C_{D1}} Z^{-1}}{1 + \dfrac{\Delta t - 2R_{D1} C_{D1}}{\Delta t + 2R_{D1} C_{D1}} Z^{-1}} \tag{2.9}$$

under Z plane, where Δt is the system sampling interval.

In Eq. (2.9), we define

$$\begin{cases} a_1 = -\dfrac{\Delta t - 2R_{D1} C_{D1}}{\Delta t + 2R_{D1} C_{D1}} \\[4mm] a_2 = -\dfrac{R_i \Delta t + R_{D1} \Delta t + 2R_i R_{D1} C_{D1}}{\Delta t + 2R_{D1} C_{D1}} \\[4mm] a_3 = -\dfrac{R_i \Delta t + R_{D1} \Delta t - 2R_i R_{D1} C_{D1}}{\Delta t + 2R_{D1} C_{D1}} \end{cases} \tag{2.10}$$

Therefore, ohmic resistance R_i can be solved from Eq. (2.10) as

$$R_i = \frac{a_3 - a_2}{1 + a_1} \tag{2.11}$$

Substituting Eq. (2.10) into Eq. (2.9) leads to

$$E_{L,k} = a_1 E_{L,k-1} + a_2 i_{L,k} + a_3 i_{L,k-1} \tag{2.12}$$

As mentioned earlier, the OCV mainly depends on battery SOC z_k, temperature T, and aging status A_{ge}. Mathematically, it can be expressed as

$$U_{oc,k} = f(z_k, T_k, A_{ge,k}) \tag{2.13}$$

Differentiating Eq. (2.13) with respect to time t results in

$$\frac{dU_{oc}}{dt} = \frac{\partial U_{oc}}{\partial z} \frac{dz}{dt} + \frac{\partial U_{oc}}{\partial T} \frac{dT}{dt} + \frac{\partial U_{oc}}{\partial A_{ge}} \frac{dA_{ge}}{dt} \tag{2.14}$$

Consider the following assumptions [11]:

1. $dz/dt \approx 0$ holds true where a small amount of energy is consumed or regained in a short time relative to the total capacity of a battery.
2. The cooling/heating system for BMSs in EVs makes the rise/decrease of battery temperature slow, namely $dT/dt \approx 0$, for normal operating conditions.
3. $dA_{ge}/dt \approx 0$ holds true over a short time since A_{ge} represents a long usage history of batteries.

Consequently, Eq. (2.14) can be simplified as $dU_{oc}/dt \approx 0$. In the discrete time domain, it can be expressed as:

$$\Delta U_{oc,k} = U_{oc,k} - U_{oc,k-1} \approx 0 \tag{2.15}$$

Substituting Eq. (2.15) into Eq. (2.12) produces

$$U_{t,k} = (1 - a_1)U_{oc,k} + a_1 U_{t,k-1} + a_2 i_{L,k} + a_3 i_{L,k-1} \tag{2.16}$$

Define the system data matrix as

$$\mathbf{\Phi}_{1,k} = \begin{bmatrix} 1 & U_{t,k-1} & i_{L,k} & i_{L,k-1} \end{bmatrix} \tag{2.17}$$

Define the parameter vector as

$$\mathbf{\theta}_{1,k} = \begin{bmatrix} (1 - a_1)U_{oc,k} & a_1 & a_2 & a_3 \end{bmatrix}^{\mathrm{T}} \tag{2.18}$$

Equation (2.16) can be written as:

$$y_k = \mathbf{\Phi}_{1,k}\mathbf{\theta}_{1,k} \tag{2.19}$$

2.4.2.3 Dual Polarization Model

When $n = 2$, the n-RC model is simplified to the dual polarization (DP) model

$$E_L(s) = -i_L(s)\left(R_i + \frac{R_{D1}}{1 + R_{D1}C_{D1}s} + \frac{R_{D2}}{1 + R_{D2}C_{D2}s} \right) \tag{2.20}$$

The corresponding transfer function is

$$G(s) = -\frac{\dfrac{1}{R_{D1}C_{D1}R_{D2}C_{D2}}(R_i R_{D1}C_{D1} + R_i R_{D2}C_{D2} + R_{D2}R_{D1}C_{D1} + R_{D1}R_{D2}C_{D2})s}{s^2 + \dfrac{R_{D1}C_{D1} + R_{D2}C_{D2}}{R_{D1}C_{D1}R_{D2}C_{D2}}s + \dfrac{1}{R_{D1}C_{D1}R_{D2}C_{D2}}}$$

$$-\frac{R_i s^2 + \dfrac{R_i + R_{D1} + R_{D2}}{R_{D1}C_{D1}R_{D2}C_{D2}}}{s^2 + \dfrac{R_{D1}C_{D1} + R_{D2}C_{D2}}{R_{D1}C_{D1}R_{D2}C_{D2}}s + \dfrac{1}{R_{D1}C_{D1}R_{D2}C_{D2}}} \tag{2.21}$$

The bilinear transformation method is employed for the discretization of Eq. (2.21), and we obtain

$$G(Z^{-1}) = \frac{b_3 + b_4 Z^{-1} + b_5 Z^{-2}}{1 - b_1 Z^{-1} - b_2 Z^{-2}} \tag{2.22}$$

where b_i $(i = 1, \cdots, 5)$ are the coefficients solved from Eq. (2.21).

Similar to the case of $n = 1$, a discretization of Eq. (2.20) is arranged as

$$U_{t,k} = (1 - b_1 - b_2)U_{oc,k} + b_1 U_{t,k-1} + b_2 U_{t,k-2} + b_3 i_{L,k} + b_4 i_{L,k-1} + b_5 i_{L,k-2} \quad (2.23)$$

Define the system data matrix as

$$\boldsymbol{\Phi}_{2,k} = \begin{bmatrix} 1 & U_{t,k-1} & U_{t,k-2} & i_{L,k} & i_{L,k-1} & i_{L,k-2} \end{bmatrix} \quad (2.24)$$

Define the parameter vector as

$$\boldsymbol{\theta}_{2,k} = \begin{bmatrix} (1 - b_1 - b_2)U_{oc,k} & b_1 & b_2 & b_3 & b_4 & b_5 \end{bmatrix}^{\mathrm{T}} \quad (2.25)$$

Equation (2.23) can be written as

$$y_k = \boldsymbol{\Phi}_{2,k}\boldsymbol{\theta}_{2,k} \quad (2.26)$$

The relationship between model parameters and coefficient b_i is:

$$
\begin{cases}
b_1 = -\dfrac{2\Delta t^2 - 8R_{D1}C_{D1}R_{D2}C_{D2}}{\Delta t^2 + 2(R_{D1}C_{D1} + R_{D2}C_{D2} + 4R_{D1}C_{D1}R_{D2}C_{D2})\Delta t} \\[3mm]
b_2 = -\dfrac{\Delta t^2 - 2(R_{D1}C_{D1} + R_{D2}C_{D2} + 4R_{D1}C_{D1}R_{D2}C_{D2})\Delta t}{\Delta t^2 + 2(R_{D1}C_{D1} + R_{D2}C_{D2} + 4R_{D1}C_{D1}R_{D2}C_{D2})\Delta t} \\[3mm]
b_3 = -\dfrac{\Delta t^2(R_{D1} + R_{D2} + R_i) + 4R_i R_{D1}C_{D1}R_{D2}C_{D2}}{\Delta t^2 + 2(R_{D1}C_{D1} + R_{D2}C_{D2} + 4R_{D1}C_{D1}R_{D2}C_{D2})\Delta t} \\[3mm]
\quad - \dfrac{2\Delta t(R_{D1}R_{D2}C_{D1} + R_{D1}R_{D2}C_{D2} + R_i R_{D1}C_{D1} + R_i R_{D2}C_{D2})}{\Delta t^2 + 2(R_{D1}C_{D1} + R_{D2}C_{D2} + 4R_{D1}C_{D1}R_{D2}C_{D2})\Delta t} \\[3mm]
b_4 = -\dfrac{4\Delta t(R_{D1} + R_{D2} + R_i) - 8R_i R_{D1}C_{D1}R_{D2}C_{D2}}{\Delta t^2 + 2(R_{D1}C_{D1} + R_{D2}C_{D2} + 4R_{D1}C_{D1}R_{D2}C_{D2})\Delta t} \\[3mm]
b_5 = -\dfrac{\Delta t^2(R_{D1} + R_{D2} + R_i) + 4R_i R_{D1}C_{D1}R_{D2}C_{D2}}{\Delta t^2 + 2(R_{D1}C_{D1} + R_{D2}C_{D2} + 4R_{D1}C_{D1}R_{D2}C_{D2})\Delta t} \\[3mm]
\quad + \dfrac{2\Delta t(R_{D1}R_{D2}C_{D1} + R_{D1}R_{D2}C_{D2} + R_i R_{D1}C_{D1} + R_i R_{D2}C_{D2})}{\Delta t^2 + 2(R_{D1}C_{D1} + R_{D2}C_{D2} + 4R_{D1}C_{D1}R_{D2}C_{D2})\Delta t}
\end{cases} \quad (2.27)
$$

2.4.2.4 *n*-RC Model

When a number of RC networks are equal to n, the model can be written in a general form as

$$E_L(s) = -i_L(s)\left(R_i + \frac{R_{D1}}{1 + R_{D1}C_{D1}s} + \frac{R_{D2}}{1 + R_{D2}C_{D2}s} + \cdots + \frac{R_{Dn}}{1 + R_{Dn}C_{Dn}s} \right) \quad (2.28)$$

The bilinear transformation method is employed for the discretization of Eq. (2.28), and we obtain

$$U_{t,k} = \left(1 - \sum_{i=1}^{n} c_i \right)U_{oc,k} + c_1 U_{t,k-1} + c_2 U_{t,k-2} + \ldots + c_n U_{t,k-n} + c_{n+1}i_{L,k}$$

$$+ c_{n+2}i_{L,k-1} + \ldots + c_{2n+1}i_{L,k-n} \quad (2.29)$$

where $c_i (i = 1, 2, \cdots, 2n + 1)$ are the parameters to be identified. Similarly, the system data parameter matrix and parameter vector are, respectively, defined as

$$\begin{cases} \boldsymbol{\Phi}_{n,k} = \begin{bmatrix} 1 & U_{t,k-1} & U_{t,k-2} & \cdots\cdots & U_{t,k-n} & i_{L,k} & i_{L,k-1} & i_{L,k-2} & \cdots\cdots & i_{L,k-n} \end{bmatrix} \\ \boldsymbol{\theta}_{n,k} = \begin{bmatrix} \left(1 - \sum\limits_{i=1}^{n} c_i\right) U_{oc,k} & c_1 & c_2 & c_3 & \cdots\cdots & c_{2n+1} \end{bmatrix}^{\mathrm{T}} \end{cases} \qquad (2.30)$$

The equation of the model can be written as

$$y_k = \boldsymbol{\Phi}_{n,k} \boldsymbol{\theta}_{n,k} \qquad (2.31)$$

It can be seen that a number of model parameters to be identified depend on the number of RC networks. The higher the number of RC networks there is, the heavier the calculation burden and the larger the memory is required to store a large amount of sampling data. Thus, it is important to select the proper number of RC networks for a battery model suitable for EV applications with an acceptable accuracy.

2.4.3 Open Circuit Voltage

Battery OCV generally refers to the potential difference between the positive electrode and the negative electrode when no external load connects to the battery and thus no external current flows through the battery. The battery OCV can go up/down immediately after the open circuit is made in the discharging/charging process. In the nth order ECM, the OCV is the static characteristic parameter which describes the relationship between the battery terminal voltage at the equilibrium state and the SOC when the battery is disconnected from a load for a certain period, where such a period of time is also known as the standing time. Depending on the types and materials of the batteries, the standing time to obtain the OCV can be from one hour to a few hours. Figure 2.4 shows the relationship between the OCV and SOC for LiBs when the standing time is one hour, where LFP has a lithium iron phosphate ($LiFePO_4$) cathode and graphite anode, LMO has a lithium manganese oxide ($LiMn_2O_4$) cathode and graphite anode, NMC has a nickel manganese cobalt oxide ($LiNiMnCoO_2$) cathode and graphite anode, and LTO has a nickel manganese cobalt oxide ($LiNiMnCoO_2$) cathode and $Li_4Ti_5O_{12}$ (titanate) anode. It can be observed that the OCV monotonically increases with the SOC. Because of this particular relationship, the OCV is widely used to estimate battery SOC. Figure 2.5 shows the OCV–SOC curves at different aging levels and operating temperatures, where the NMC battery has been tested for 600 cycles. The OCV–SOC curves for the particular cycles are plotted in Figure 2.5a. The partially magnified view in Figure 2.5a shows that the OCV increases with the number of cycles. The OCV–SOC curves for the fresh cell (cycle number is zero, i.e. cyc00) at temperatures of 10, 25, and 40 °C are plotted in Figure 2.5b. The partially magnified view in Figure 2.5b shows that the OCV decreases with temperature rise.

The OCV–SOC curves of a LFP battery during the charging/discharging process at different temperatures of 10, 25, and 40 °C are plotted in Figure 2.6. The magnified

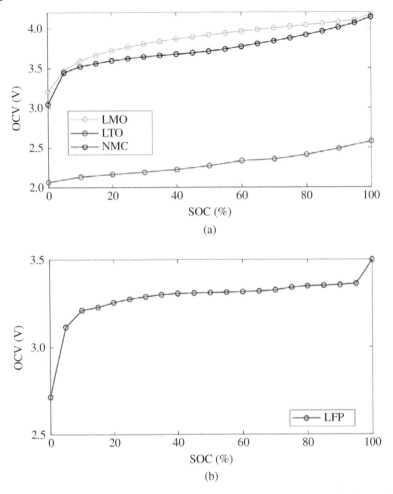

Figure 2.4 OCV–SOC relationships for four types of LiB materials: (a) LMO, LTO, and NMC; (b) LFP [12].

views in Figure 2.6a–c show that the OCVs in the charging process do not retrace back the OCVs in the discharging process and furthermore the OCVs in the charging process are generally larger than the OCVs in the discharging process at the corresponding temperatures. Such differences are called the hysteresis characteristic for LiBs, which is similar to the vehicle tire elastic hysteresis where loading and unloading trace out a different path.

Take a LFP battery as an example to explain the hysteresis phenomenon. There are several possible thermodynamic equilibrium potentials at the same SOC of a cell. In the cell, a positive electrode with LFP as active material exhibits a hysteretic phenomenon due to the existence of a lithium rich and lithium deficient phase within an active particle [13], which are termed the path dependent shrinking core model. According to this model, a shrinking particle core of Li_yFePO_4 and a growing outer crust of $Li_{(1-x)}FePO_4$ occur during discharging while a shrinking core of $Li_{(1-x)}FePO_4$ and a growing crust of Li_yFePO_4 occur during charging (considering mole fractions x and y are close to zero). Depending on the two-phase particle composition, the chemical potential of the particle, namely

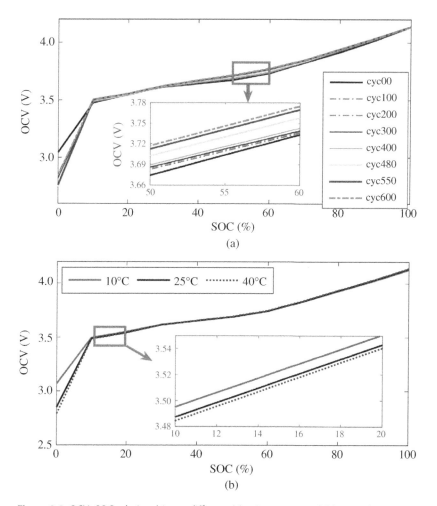

Figure 2.5 OCV–SOC relationships at different (a) aging status and (b) operating temperatures.

open circuit potential, can be different at the same SOC. The charging and discharging of a LiB is essential to intercalation and de-intercalation of Li^+ from the $LiFePO_4$ crystals of olivine structure, where the volume and structure of the crystals will change. This phase transition produces some tension or pressure, which is stored inside the crystal in the form of "elastic-plastic" energy. This is reflected as the battery voltage hysteresis phenomenon in the charging and discharging processes.

According to the voltage hysteresis characteristic of an LFP battery, the hysteresis voltage has the following characteristics:

1. When the charging or discharging of a battery stops and the load current becomes zero, the hysteresis voltage does not disappear immediately. It remains the same for some time and then gradually reduces, but it does not necessarily reduce to zero.
2. The hysteresis voltage is decided by the last charge and discharge characteristics and is limited to the maximum and minimum hysteresis voltage dependent on electrochemical reactions inside the battery.

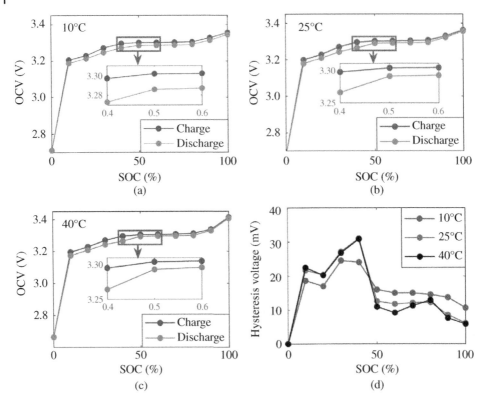

Figure 2.6 OCV–SOC relationships during charging and discharging processes at different temperatures: (a) 10 °C; (b) 25 °C; (c) 40 °C, and (d) their hysteresis voltage differences.

Figure 2.6d shows that the hysteresis voltage differences of the LFP battery vary with the similar trend at different temperatures, the maximum voltage difference up to 30 mV can be observed at the temperatures of 25 and 40 °C when the SOC is 40%. Since the OCV versus the SOC has hysteresis characteristics, when the OCV is used to estimate the SOC it will cause the SOC estimation error.

2.4.4 Polarization Characteristics

Electrode polarization is very common in a battery. When no current flows through the battery for a certain period of time, the battery is considered to be in the equilibrium state and the corresponding electrode potential is called the equilibrium electrode potential. When current flows through the battery, the electrode potential deviates from the equilibrium potential; this phenomenon is called electrode polarization or overpotential. The larger the current flowing through electrode, the larger the deviation from the equilibrium electrode potential. There are three types of polarization:

1. Electrochemical polarization is caused by various types of redox reaction on electrodes.
2. Concentration polarization is caused by the fact that the reactants consumed on the electrode surface cannot be supplemented from the electrolyte in time (or some reaction products cannot be evacuated in time).

3. Ohmic polarization represents the resistances of electrolyte, electrode material, and conductive material and their contact resistances.

The internal resistance of a battery is generally the sum of ohmic polarization resistance, electrochemical polarization resistance, and concentration polarization resistance.

2.5 Experiments

Figure 2.7 shows a battery test bench. It consists of two Arbin BT2000 cyclers, four thermal chambers to regulate operating temperatures of the batteries, a Zahner IM6 electrochemical workstation to test electrochemical impedance spectroscopy (EIS) of batteries and a host computer to program and store experimental data. For the two Arbin cyclers, one cycler has 16 dependent channels which allows to load the current or power profiles on LiB cells with voltage limits of 0–5 V and current limits of ± 100 A; the other cycler is a battery pack cycler which allows to load the current or power profiles on the LiB packs with voltage limits of 0–60 V and current limits of ± 300 A. The accuracy of the current and voltage measurement of the Arbin test systems is within 0.05%. The measured data is sent to the host computer through TCP/IP ports. The LiB cells are connected to the Arbin BT2000 cycler and then placed inside the thermal chamber to perform the tests at the desired temperature. The temperature range of the thermal chambers is between -55 and $120\,°C$.

The three types of LiB cells: LFP, NMC, and LMO are tested to obtain the experimental data. The specifications of these cells are listed in Table 2.1. The test schedules shown in Figure 2.8 are designed to generate rich excitations for these three types of cells. A static capacity test, a Columbic efficiency test, a hybrid pulse test, an OCV versus SOC test, and loading profiles test are consecutively conducted.

The static capacity test is to obtain the cell's maximum available capacity at its current state, which could be different from its nominal capacity. The charge–discharge

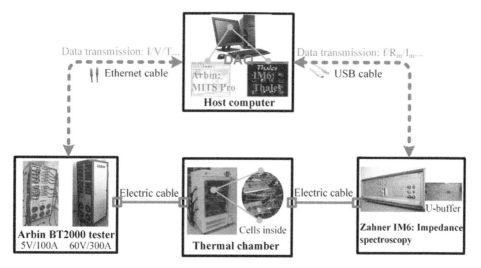

Figure 2.7 Battery test bench.

Table 2.1 Specifications of three types of battery cells.

Type	Rated capacity (Ah)	Rated voltage (V)	Upper cut-off voltage (V)	Lower cut-off voltage (V)	Nominal current (A)
NMC	32	3.60	4.05	3.0	16
LMO	35	3.70	4.20	3.0	12
LFP	20	3.20	3.65	2.50	10

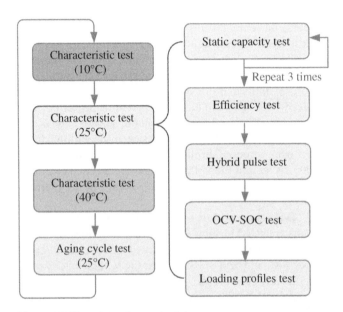

Figure 2.8 Flow chart of test schedules.

Coulomb efficiency test is to obtain the Coulomb efficiency of the batteries under different operation currents for battery model compensation. Instead of the traditional hybrid pulse power characterization (HPPC) test which only uses $1C^1$ discharge rate and 0.75C charge rate, the specific HPPC (SHPPC) test is adopted to improve the applicability of the battery model under a wide operation range of dynamic driving cycles. The SHPPC is a sequence of the pulse cycles consisting of four different charge–discharge currents, considering that the typical current ranges of the battery in EVs are between 0 and 3C. The sampling interval in the experiments is one second. Figure 2.9 shows the experimental data of the current, voltage, and SOC versus time for the C/NMC LiB cell under the SHPPC test. The OCV measurement requires high precision, especially for the LFP cell because of its flat voltage characteristics in the

---

1 In the battery systems, the C rate represents the battery charge/discharge rate (in amperes) relative to its full capacity. For a LiB with a nominal capacity of 35 Ah, a discharge current at 1C rate means that the discharge current is 35 A.

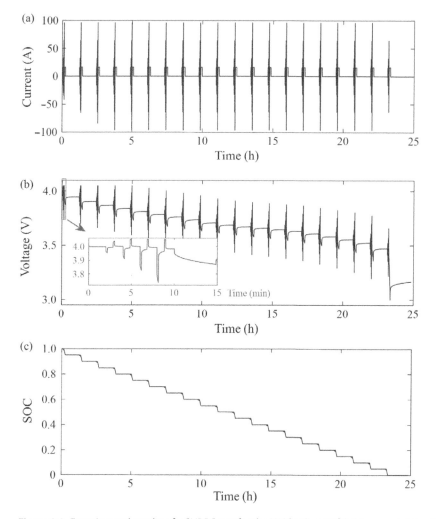

Figure 2.9 Experimental results of a SHPPC test for the NMC cell at 25 °C: (a) current; (b) voltage; (c) SOC.

middle range of the SOC. The curve of the OCV versus the SOC is required to estimate the SOC and other battery states.

This curve can be obtained by performing the following tests: (1) Fully charge the cell with the constant current constant voltage (CC/CV) charging method, where the constant current of C/3 standard current is used to charge the cell to the upper cut-off voltage until the current tapers to the cut-off current of C/20. Then, take a rest for five hours to complete the process of depolarization. (2) Discharge the cell with 5% of the maximum available capacity with the standard current until its lower cut-off voltage is reached, and leave the cell in an open-circuit condition for five hours to obtain the OCVs. (3) Repeat step (2) to obtain all OCVs in the discharging process. (4) Charge the cell for

5% of the maximum available capacity until the charging current reduces to C/20, and leave the cell in an open-circuit condition for five hours to obtain the OCVs. (5) Repeat step (4) to obtain all the OCVs in the charging process. (6) Obtain the OCVs in both the discharging process and the charging process.

The Urban Dynamometer Driving Schedule (UDDS) and Federal Urban Driving Schedule (FUDS) tests are conducted to validate the accuracy, robustness and reliability of battery models. The current, voltage and SOC profiles of the NMC cell under the UDDS test are plotted in Figure 2.10, where the initial SOC is set to 0.9. The profiles of the LMO cell under the FUDS test are plotted in Figure 2.11, where the initial SOC is set to 1.0. In these two cases, the "true" SOC is calculated by using the Coulomb counting method compensated by the Coulomb efficiency based on the experimental data obtained from Arbin test equipment.

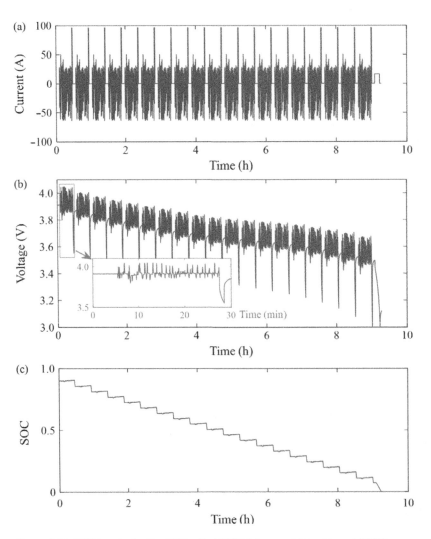

Figure 2.10 UDDS cycles for the NMC cell at 25 °C: (a) current; (b) voltage; (c) SOC.

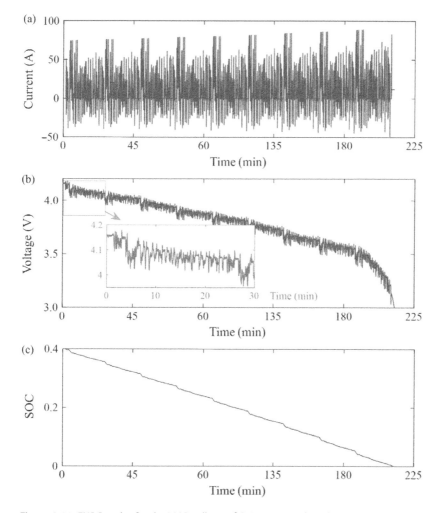

Figure 2.11 FUDS cycles for the LMO cell at 25 °C: (a) current; (b) voltage; (c) SOC.

2.6 Parameter Identification Methods

Depending on the use of experiment data offline or online, parameter identification methods can be divided into the offline parameter identification method (OFFPIM) and online parameter identification method (ONPIM).

2.6.1 Offline Parameter Identification Method

The OFFPIM updates and calibrates battery model parameters using voltage, current, and temperature stored in BMSs for EVs. It can also determine the model order suitable for a particular application to achieve the balance between model calculation burden and accuracy, which can be implemented in the following five steps [14].

Step 1: Extract archived operation data. It first extracts the measurement of battery current, voltage, and SOC from the archived operation data stored in a BMS. With

5% of the SOC interval of the adjacent parameter group and the whole SOC range, we can calculate the number N of the data group for parameter identification, where m denotes the order of the data group, $m = 1, 2, \ldots, N$.

Step 2: *Initialization.* It initializes α and λ. α represents the SOC range of each group; its initial value is set to $[-0.5\%, 0.5\%]$. λ represents the allowable times for enlarging the data range during parameter identification; its initial value is 1. For instance, if the parameter identification result under $\alpha = [-0.5\%, 0.5\%]$ cannot meet the demand of the fitting factor, the SOC range α will be enlarged to $[-0.5\% - \lambda \times 0.1\%, 0.5\% + \lambda \times 0.1\%]$, where λ will be increased by 1 later.

Step 3: *Parameter identification.* By providing a number of RC networks or the order for the battery model n ($n = 0, 1, 2, 3, 4, 5, \ldots$), we can identify the model parameters under different numbers of RC networks. Then, the coefficient of determination R^2 is used to statistically measure the closeness between the model and the data. We set the minimal value of R^2 as 0.98. If $R^2 \leq 0.98$, we will enlarge the SOC range. However, the maximum allowable times λ is 5, namely the widest SOC range is $[-1\%, 1\%]$. If all the R^2 in the five times are less than 0.98, the group of parameters with the largest R^2 will be selected as the battery model parameters at this SOC. Until the battery model parameters for all the SOC points are identified, namely $m = N$, step 3 will transfer to the next step.

Step 4: *Model estimation errors.* With the model parameters under different numbers of RC networks, we can calculate the model estimation errors. It is worth noting that the battery model with more than 30 mV error would be removed as an unfeasible model. The remaining battery models will be ready for the order evaluation.

Step 5: *Determine model order and update the model parameters.* Based on the Akira Information Criterion (AIC), the accuracy and complexity of the model are evaluated for different model orders (except the inefficient order model). As $n \ll L_D$ and $\hat{s}_k^2 < 1\ V^2$, AIC can be simplified to

$$\text{AIC} = 2 \log \hat{s}_k^2 + 2n \tag{2.32}$$

where $\hat{s}_k^2$ represents the average of the residual squares under the optimal parameters:

$$\hat{s}_k^2 = \frac{1}{L_D} \sum_{k=1}^{L_D} (y_k - \hat{y}_k)^2 \tag{2.33}$$

where L_D is the length of the data, and $\hat{y}_k$ is the estimated terminal voltage.

Figure 2.12 shows the flow chart to implement the OFFPIM.

Many methods can be used to identify the model parameters in step 3, such as the least squares (LS) method, curve fitting method, constrained nonlinear minimization method, minimax optimization method, and genetic algorithm (GA). For linear systems described in Section 2.4, LS is an efficient algorithm. Take the Thevenin model (or the first-order RC network model) using the LS method to identify its parameters as an example. Its system information matrix is expressed as

$$\Phi = \begin{bmatrix} O_1 & U_t^{0:n_d-1} & I_L^{1:n_d} & I_L^{0:n_d-1} \end{bmatrix} \tag{2.34}$$

where n_d is the length of the data for the battery current (I_L)

$$n_d = \text{length}(I_L) \tag{2.35}$$

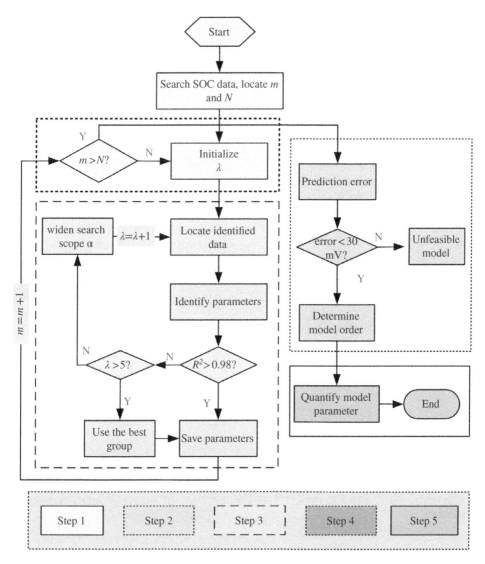

Figure 2.12 Flow chart of the offline parameter identification method.

$\mathbf{O}_1$ is an $n_d \times 1$ vector and all values are 1, $\mathbf{U}_t^{0:n_d-1}$, $\mathbf{I}_L^{1:n_d}$, and $\mathbf{I}_L^{0:n_d-1}$ are $n_d \times 1$ vectors which are, respectively, denoted by

$$\begin{cases} \mathbf{U}_t^{0:n_d-1} = \left[U_{t,0} \ U_{t,1} \ \cdots \ U_{t,n_d-1} \right]^{\mathrm{T}} \\ \mathbf{I}_L^{1:n_d} = \left[i_{L,1} \ i_{L,2} \ \cdots \ i_{L,n_d} \right]^{\mathrm{T}} \\ \mathbf{I}_L^{0:n_d-1} = \left[i_{L,0} \ i_{L,1} \ i_{L,2} \ \cdots \ i_{L,n_d-1} \right]^{\mathrm{T}} \end{cases} \tag{2.36}$$

$\mathbf{Y}$ is an $n_d \times 1$ vector of measured voltage:

$$\mathbf{Y} = \left[U_{t,1} \ U_{t,2} \ \cdots \ U_{t,n_d} \right]^{\mathrm{T}} \tag{2.37}$$

We can obtain the model parameters by solving

$$Y = \Phi\theta \tag{2.38}$$

θ can be solved as:

$$\hat{\theta} = (\Phi^T\Phi)^{-1}\Phi^T Y \tag{2.39}$$

and it is composed of:

$$\theta = \begin{bmatrix} (1-a_1)U_{oc} & a_1 & a_2 & a_3 \end{bmatrix}^T \tag{2.40}$$

It is noted that the data matrix Φ is slightly different from that in Eq. (2.17) while the parameter vector θ is the same as in Eq. (2.18).

The optimization method can also be used to identify the model parameters. Generally, the optimization method should have three aspects of information including an objective function (a loss function or cost function), optimal arguments, and constraints. Take the Thevenin model (or the first-order RC network model) as an example again. They can be described by

$$\text{Objective function}: \quad f = \min\left\{\sum_{k=1}^{n_d}(y_k - U_t)^2\right\} \tag{2.41}$$

$$\text{Optimal arguments}: \quad x = \begin{bmatrix} U_{oc} & R_i & R_{D1} & \tau_1 \end{bmatrix} \tag{2.42}$$

$$\text{Subject to}: \quad \begin{cases} R_{i,L} \leq R_i \leq R_{i,U} \\ R_{D1,L} \leq R_{D1} \leq R_{D1,U} \\ U_{oc,L} \leq U_{oc} \leq U_{oc,U} \\ \tau_{i,L} \leq \tau_1 \leq \tau_{1,U} \end{cases} \tag{2.43}$$

where $R_{o,L}$ and $R_{o,U}$ denote the lower and upper bounds of ohmic resistance, respectively, $R_{D1,L}$ and $R_{D1,U}$ denote the lower and upper bounds of polarization resistance, respectively, and $\tau_{1,L}$ and $\tau_{1,U}$ denote the lower and upper bounds of the time constant, respectively, which are the products of the polarization capacitor and polarization resistance.

2.6.2 Online Parameter Identification Method

The ONPIM updates the model parameters by using the real-time measurement of battery current, voltage, and temperature. The ONPIM is also called the online data-driven parameter identification method. The LS method used in the OFFPIM is a batch algorithm which requires all the system input and output data for model parameter identification so that the residual between the model output and system output data is minimized. Different from the LS method, the recursive least squares (RLS) method is used in the ONPIM. The RLS method is a real-time system identification method based on adaptive filtering theory [15].

The ONPIM recursively updates the model parameters as new observations arrive, the satisfactory parameters are obtained when the RLS is converged. For the case where the battery model and parameters are susceptible to uncertainties in real EV applications, the RLS method regularly updates the model parameters to capture the

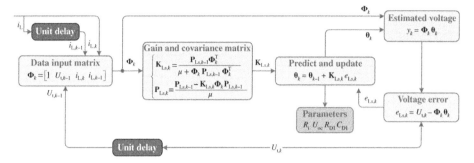

Figure 2.13 Flowchart of online parameter identification method.

real-time characteristics of the battery. For the battery model parameters, since they are changing slowly, the RLS method with a forgetting factor (RLS-FF) is adopted. It provides more weightage on new measurement data than old measurement data. For the system expressed by

$$y_k = \mathbf{\Phi}_k \mathbf{\theta}_k + e_{\mathrm{Ls},k} \tag{2.44}$$

where $e_{\mathrm{Ls},k}$ is the zero-mean white noise, the symbol is defined as the same as those in (2.30). Then, the RLS-FF method can be implemented by

$$\begin{cases} \mathbf{K}_{\mathrm{Ls},k} = \mathbf{P}_{\mathrm{Ls},k-1} \mathbf{\Phi}_k^{\mathrm{T}} [\mathbf{\Phi}_k \mathbf{P}_{\mathrm{Ls},k-1} \mathbf{\Phi}_k^{\mathrm{T}} + \mu]^{-1} \\ \hat{\mathbf{\theta}}_k = \hat{\mathbf{\theta}}_{k-1} + \mathbf{K}_{\mathrm{Ls},k} [y_k - \mathbf{\Phi}_k \hat{\mathbf{\theta}}_{k-1}] \\ \mathbf{P}_{\mathrm{Ls},k} = \frac{1}{\mu} [\mathbf{I} - \mathbf{K}_{\mathrm{Ls},k} \mathbf{\Phi}_k] \mathbf{P}_{\mathrm{Ls},k-1} \end{cases} \tag{2.45}$$

where μ is a forgetting factor; Eq. (2.45) degenerates into the RLS method when $\mu = 1$. $\mathbf{K}_{\mathrm{Ls},k}$ is the gain and $\mathbf{P}_{\mathrm{Ls},k}$ is the error covariance matrix for the estimated states. With Eqs. (2.17), (2.18), and (2.45), the ONPIM based on the RLS-FF method is established to identify the parameters of the Thevenin model (or the first-order RC network model); the flowchart of its implementation is shown in Figure 2.13. The ONPIM can also identify the model parameters for different numbers of RC networks.

2.7   Case Study

2.7.1   Testing Data

The experimental data of the NMC and LMO cells at a temperature of 25 °C are used in the following parameter identification and validation process, where the specifications of these two types of battery cells can be found in Table 2.1. If the experimental data at temperatures other than 25 °C are used in the process, it will be marked specifically.

2.7.2   Case One – OFFPIM Application

The OFFPIM based on the LS method will be used to select the model structure (e.g. model orders) and identify parameters of battery ECMs and evaluate their suitability

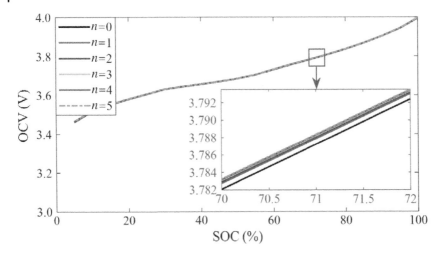

Figure 2.14 OCVs at different orders of ECMs identified by the OFFPIM for the NMC cell.

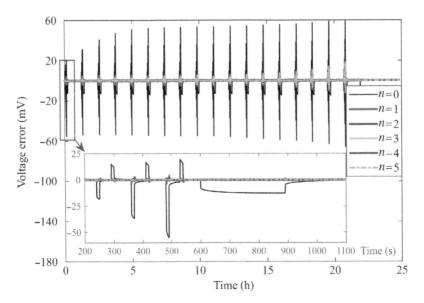

Figure 2.15 Comparison of OCV errors for different orders of ECM for the NMC cell.

for EV applications. The OCV is one of the most important characteristics of LiBs. Figures 2.14 and 2.15 are the identified OCVs by using ECMs with orders of 0–5 and the comparison of their estimation errors under the SHPPC test data of the NMC cell shown in Figure 2.9. It can be seen from the magnified part that the OCVs of the ECMs with orders of 1–5 are almost the same, but they are very different from that of the ECM with order of 0.

Figure 2.15 shows the error comparison of the OCVs for six ECMs with the standard OCV which is calibrated in the experiment. It can be seen from the magnified part that all of the maximum OCV errors for the ECMs with orders of 1–5, namely the ECMs

Table 2.2 Comparison of ECMs at different orders of RC networks under SHPPC testing data using the OFFPIM.

Order	R^2	Maximum error (mV)	Average error (mV)	Time for look-up table (s)	Time for curve fitting (s)
$n = 0$	0.926	63.789	1.701	19.61	0.61
$n = 1$	0.991	19.002	0.116	36.28	1.12
$n = 2$	0.992	18.508	0.114	71.08	3.56
$n = 3$	0.992	18.131	0.112	94.32	7.13
$n = 4$	0.993	19.115	0.112	119.53	15.32
$n = 5$	0.991	19.079	0.111	147.63	27.21

with RC networks, are less than the pre-set error threshold of 30 mV while the maximum OCV error (63 mV) for the ECM with the order of 0, namely the ECM without any RC network or R_{int} model, is larger than 30 mV.

Thus, the OCV errors of ECMs with RC networks meet the accuracy requirement which shows that the RC networks can improve the performances of ECMs.

To evaluate the accuracy and calculation times of six ECMs, Table 2.2 shows the comparison for the ECMs with different orders, where R^2 takes the minimum value of N (in this case it is 19) groups of identification parameters and its threshold value sets to 0.98. It can be seen that only the R_{int} model has R^2 smaller than its threshold. Hence, R_{int} model cannot meet the accuracy requirement when it is used to describe dynamic characteristics of LiBs. It can also demonstrate that the highest estimation accuracy occurs at the third order of the ECM rather than the fifth order of the ECM, indicating that the higher order of ECM may not provide higher estimation accuracy of the ECM. In the modeling process, the look-up table or the curve fitting method is used to describe the relationship between OCVs and SOCs. The look-up table provides a set of OCVs corresponding to a set of SOCs for 19 sets of the identified model parameters. The curve fitting method uses the sixth order polynomial to fit the OCVs versus the SOCs for the same sets of data as the look-up table. It can be observed that the look-up table takes longer than the curve fitting method and the maximum error of the R_{int} model is 63.789 mV, which is much larger than 30 mV. If an error compensation function is added to the R_{int} model, its estimation accuracy is improved with a maximum error only of 21 mV. The results are shown in Figure 2.16.

Table 2.3 provides the AIC values to show the degree of the trade-offs between the accuracy and complexity of the ECMs with their different orders. The ECM with the minimum AIC is regarded as the best model. According to Table 2.3, one RC network model or Thevenin model with the lowest AIC value of 2.73 is the best model for the NMC cell.

The ECMs at different orders of RC networks are identified under the SHPPC testing data. The currents in the UDDS testing data as shown in Figure 2.10 are substituted into these ECMs to estimate terminal voltages which are then compared with those terminal voltages in the UDDS testing data. Their comparison results in terms of the maximum errors, average errors, AICs, and calculation times for the ECMs at different orders of RC networks are listed in Table 2.4, where the ECM without any RC network (namely

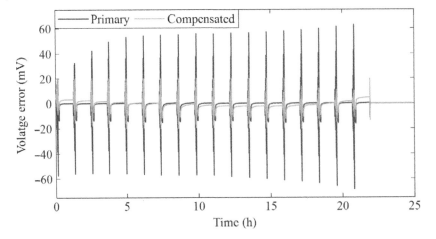

Figure 2.16 Results of voltage estimation error considering the error compensation function for the R_{int} model.

Table 2.3 AICs for the NMC cell using the OFFPIM.

Order	$n = 0$	$n = 1$	$n = 2$	$n = 3$	$n = 4$	$n = 5$
AIC	12.01	2.73	10.46	14.68	17.79	20.53

Table 2.4 Statistical results of voltage estimation error with the UDDS testing data using the OFFPIM.

Order	Maximum error (mV)	Average error (mV)	AIC	Calculation time (s)
$n = 0$	29.897	2.070	11.57	0.47
$n = 1$	27.966	0.828	6.31	1.69
$n = 2$	27.694	0.796	8.21	3.06
$n = 3$	27.791	0.775	10.16	5.13
$n = 4$	27.913	0.760	12.12	8.32
$n = 5$	27.762	0.748	14.08	11.21

$n = 0$) has been enhanced by the error compensation function. It can be seen that all the errors of the estimated terminal voltages are less than 30 mV for the ECMs with orders from 0 to 5. The results suggest that the battery model parameters identified by the OFFPIM have good adaptability to different battery operation conditions.

2.7.3 Case Two – ONPIM Application

The ONPIM based on the RLS method is used to determine the model orders and identify the model parameters under the same experimental data of the battery cell as those

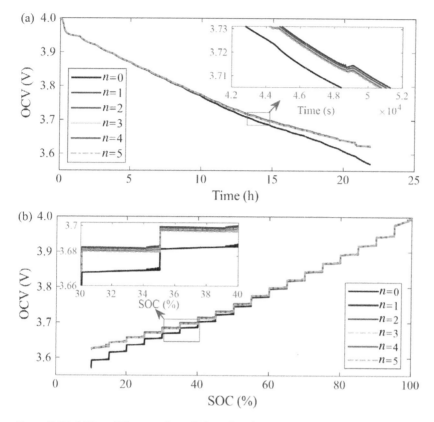

Figure 2.17 OCVs at different orders of ECMs identified by ONPIM for the NMC cell: (a) OCV versus time; (b) OCV versus SOC.

in Section 2.7.2. Figure 2.17 shows the identified OCVs for the ECMs with different orders. From observation of Figure 2.17, three conclusions are reached:

1. The OCVs for all ECMs with orders of 0–5 tend to be similar from the beginning to about half of the total operation time or from SOCs of 50 to 100%.
2. The OCV for the ECM without any RC network (namely $n=0$) diverged from the other OCVs for the ECMs with orders of 1–5 from about half of the total operation time to the end or from SOCs of 50% to 0%.
3. The OCVs for the UDDS test at the SOC interval of 5% remain the same as those for the SHPPC test at the SOC interval of 5%, indicating that the relationship between the OCVs and the SOCs using ONPIM is stable regardless of different battery operation conditions.

Figure 2.18 shows the estimated OCV errors between the model and the experimental data for the ECMs with different orders of RC networks. It can be seen that all the errors are less than ±50 mV.

The OCVs quickly converge with increase of testing data. After convergence, the errors are all within 15 mV except for the ECM without any RC network, which indicates the effectiveness of the ONPIM. Table 2.5 shows that all the maximum errors for the ECMs with orders of 1–5 ($n>0$) are within 30 mV under SHPPC testing data

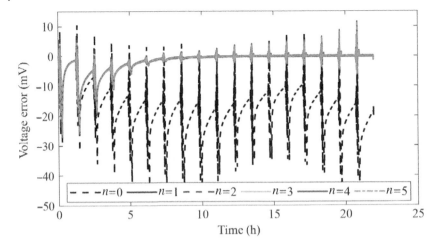

Figure 2.18 Estimated OCV errors for ECMs with different orders.

Table 2.5 Statistical results of voltage estimation error under SHPPC testing data using the ONPIM.

Order	Maximum error (mV)	Average error (mV)	AIC	Calculation time (s)
$n=0$	46.588	15.459	17.16	0.61
$n=1$	27.213	1.521	12.51	1.04
$n=2$	26.892	1.516	14.51	1.36
$n=3$	26.689	1.514	16.51	1.51
$n=4$	26.483	1.514	18.50	1.82
$n=5$	26.345	1.514	20.50	2.45

and the average error is less than 2 mV except for the error of the ECM without any RC network. According to the AIC, the ECM with the first-order RC network ($n=1$) is the best model and its computation time is only 1.04 seconds. Table 2.6 shows the validation results under the UDDS testing data.

2.7.4 Discussions

According to AIC, the ECM with the first-order RC network, namely $n=1$, is more suitable for the NMC cell regardless of the OFFPIM or the ONPIM. From Tables 2.4 and 2.6, it can be seen that the model estimation accuracy of the OFFPIM is slightly higher than that of the ONPIM for one set of experimental data. Since battery model parameters change slowly with cyclic operation of batteries in EVs and the ONPIM can capture the changes of battery model parameters in real-time, the ONPIM will be better than the OFFPIM in terms of the applicability and adaptability of battery models in EVs. Furthermore, an imbedded controller can implement the ONPIM much more easily than the OFFPIM because the ONPIM requires less computational time and less memory.

Table 2.6 Statistical results of voltage estimation error under UDDS testing data using the ONPIM.

Order	Maximum error (mV)	Average error (mV)	AIC	Calculation time (s)
$n=0$	30.902	7.249	10.38	0.32
$n=1$	29.896	1.892	8.51	0.87
$n=2$	29.205	1.885	10.50	1.15
$n=3$	28.285	1.886	12.50	1.31
$n=4$	27.534	1.888	14.49	1.42
$n=5$	26.828	1.891	16.48	1.67

2.8 Model Uncertainties

Batteries in EVs are operating in complex and highly dynamic environments, leading to many uncertain factors to affect the accuracy of ECMs. In the following, the OFFPIM and the ONPIM will be analyzed and evaluated by considering these uncertain factors, such as battery aging, battery type, and battery temperature.

2.8.1 Battery Aging

Battery aging affects the internal chemical reactions and thermal process of LiBs which will, in turn, change the parameters of battery ECMs. The ECM with constant parameters cannot reflect such changes. The ECM identified by using SHPPC testing data at the fresh NMC cell with a capacity of 31.8 Ah in Section 2.7.2 is applied to the aged NMC cell with a capacity of 25 Ah to investigate the influence of aging status on ECM accuracy and evaluate the performances of the OFFPIM and the ONPIM.

The ECM identified by the OFFPIM is used to estimate the terminal voltages of the aged cell under UDDS testing data. Figure 2.19 shows the voltage estimation errors between the model and experimental data. It can be seen that large errors are created from all six ECMs for the aged cell.

Table 2.7 lists the maximum errors, average errors, and root mean squared errors (RMSEs) for the ECMs with different orders of RC networks. The comparison between Tables 2.7 and 2.4 shows that the maximum errors for the aged battery are about five times larger than those for the fresh battery and the accuracies of all six ECMs deteriorate due to battery aging. This is due to the fact that the OFFPIM has no mechanism to update model parameters in real-time to capture the change of battery model parameters. Thus, the ECMs identified by the OFFPIM cannot predict terminal voltage accurately during the full service life of LiBs.

On the other hand, the ONPIM provides the mechanism to update and calibrate the model parameters adaptive to the UDDS testing data for the aged cell. Figure 2.20 plots the voltage errors between the ECMs identified by the ONPIM and experimental data and these errors are much smaller than those in Figure 2.19. Table 2.8 lists the voltage prediction errors for all six ECMs with the orders of 0–5. They all have errors close to those in Table 2.4. It shows that the ONPIM can effectively solve the problem of accuracy reduction caused by battery aging and achieve higher model accuracy than the OFFPIM.

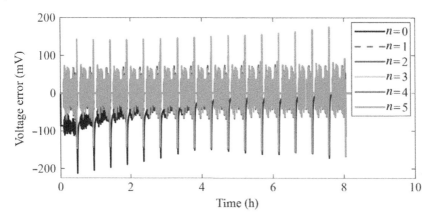

Figure 2.19 Voltage predication errors under UDDS testing data of the aged cell using the OFFPIM.

Table 2.7 Statistical results of voltage predication errors for the aged cell under UDDS testing data using the OFFPIM.

Order	Maximum error (mV)	Average error (mV)	RMSE (mV)
$n=0$	212.47	30.94	32.40
$n=1$	167.42	6.94	10.06
$n=2$	173.02	6.63	10.08
$n=3$	174.56	6.85	10.26
$n=4$	175.36	6.93	10.32
$n=5$	175.69	6.97	10.37

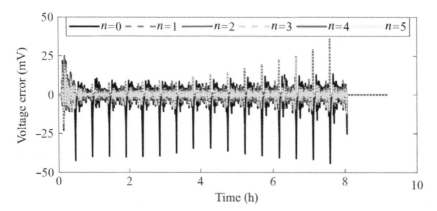

Figure 2.20 Voltage prediction errors for the aged cell under UDDS testing data using the ONPIM.

Table 2.8 Voltage prediction errors for the aging cell under UDDS testing data using the ONPIM.

Order	Maximum error (mV)	Average error (mV)	RMSE (mV)
$n=0$	43.93	2.75	4.73
$n=1$	29.40	0.78	1.53
$n=2$	29.58	0.60	1.42
$n=3$	29.52	0.56	1.37
$n=4$	30.54	0.53	1.33
$n=5$	30.45	0.51	1.29

2.8.2 Battery Type

The experimental data of the LMO cell in Figure 2.11 are used to investigate the suitability of both the OFFPIM and the ONPIM to another type of LiBs. Figure 2.21 shows the OCVs versus the SOCs for the ECMs with different orders of RC networks using the OFFPIM. It can be seen from the magnified part that the OCVs for the ECMs with the orders of 1–5 are all close to each other while the OCV for the ECM without any RC network deviates from all other OCVs, which is similar to those for the NMC cell in Figure 2.14.

Figure 2.22 shows the terminal voltage errors between the model prediction and experimental data. It should be noted that the error compensation function is added to the ECM without any RC network. The detailed statistical comparison results are shown in Table 2.9. It indicates that the OFFPIM can accurately predict terminal voltage with all errors less than 30 mV. According to AIC, the ECM with the first order of RC network is the best model with the average error and the RMSE less than 3 mV. Figure 2.23 shows the OCVs versus the SOCs for the ECMs with different orders of

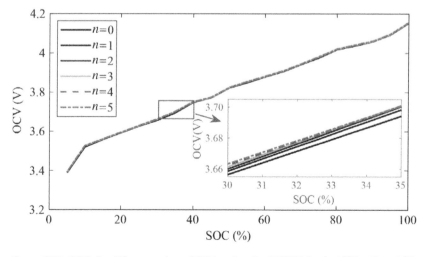

Figure 2.21 OCVs for different orders of ECMs using the OFFPIM for the LMO cell at 40 °C.

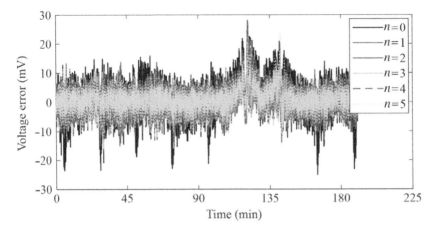

Figure 2.22 Voltage prediction errors under FUDS testing data using the OFFPIM for the LMO cell at 40 °C.

Table 2.9 Statistical results for the LMO cell using the OFFPIM.

Order	Maximum error (mV)	Average error (mV)	AIC	RMSE (mV)
$n = 0$	27.64	5.00	10.12	4.45
$n = 1$	28.17	2.86	9.61	2.89
$n = 2$	27.70	2.62	11.30	2.70
$n = 3$	27.27	2.55	13.20	2.63
$n = 4$	27.60	2.56	15.23	2.66
$n = 5$	24.75	2.51	17.04	2.47

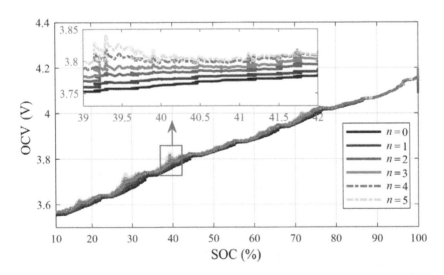

Figure 2.23 OCVs for different orders of ECMs using the ONPIM for the LMO cell at 40 °C.

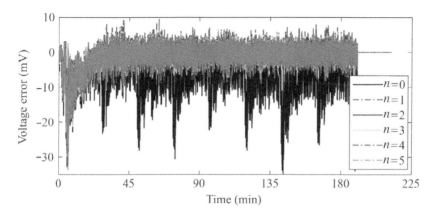

Figure 2.24 Voltage prediction errors under FUDS testing data using the ONPIM for the LMO cell at 40 °C.

Table 2.10 Statistical results for the LMO cell using the ONPIM.

Order	Maximum error (mV)	Average error (mV)	AIC	RMSE (mV)
$n=0$	34.78	6.70	10.69	5.68
$n=1$	30.38	2.22	9.11	2.82
$n=2$	31.77	2.06	10.93	2.74
$n=3$	31.44	2.00	12.84	2.70
$n=4$	32.03	1.96	14.78	2.65
$n=5$	33.71	1.94	16.73	2.62

RC networks using the ONPIM. It can be observed from the magnified part that the higher model order causes the poor stability of the OCV prediction. Figure 2.24 shows the terminal voltage errors between the model prediction and the experimental data. Their detailed statistical comparison results are shown in Table 2.10. It shows that the ONPIM has strong adaptability to another type of LiBs. Moreover, the AIC values indicate that the ECM with the first order of RC network is most suitable for the LMO cell. Therefore, the OFFPIM and the ONPIM both can accurately predict the OCVs for the LiBs with different types of batteries.

2.8.3 Battery Temperature

Battery temperature has a strong influence on the performance of LiBs. The ECM identified using the FUDS testing data at 40 °C is used to predict the terminal voltage of the LMO cell and the predicted terminal voltages at 40 °C are compared with those obtained from the experimental data at 10 °C. Figure 2.25 shows the voltage error between the model prediction and experimental data.

Table 2.11 provides the statistical comparison results of voltage prediction error for the ECMs with different orders of RC networks using the OFFPIM. It can be seen that

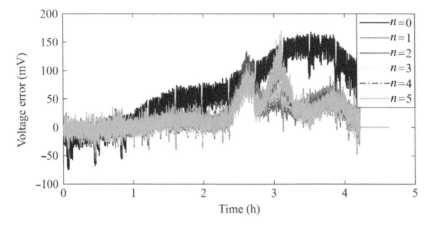

Figure 2.25 Voltage prediction errors at 10 °C when the model parameters are identified at 40 °C for the LMO cell.

Table 2.11 Statistical results of voltage prediction error using the OFFPIM at 10 °C.

Order	Maximum error (mV)	Average error (mV)	RMSE (mV)
$n = 0$	167.26	62.07	50.29
$n = 1$	126.91	24.36	24.61
$n = 2$	128.00	21.72	22.67
$n = 3$	127.32	21.10	22.85
$n = 4$	134.29	23.59	27.01
$n = 5$	170.62	24.39	26.92

there is an increase in the maximum errors, average errors, and RMSEs of terminal voltages. The maximum errors and the average errors are much larger than those in Table 2.4 and the RMSEs are above 20 mV. For comparison, the ONPIM is used to identify the model parameters and predict the terminal voltage at 10 °C. The statistical comparison results are shown in Table 2.12. It shows that the ECM identified by the ONPIM can predict terminal voltage accurately under different battery operating temperatures. This is due to the fact that the ECM identified by the ONPIM can update model parameters in real-time.

2.9 Other Battery Models

There are many other battery models commonly used in BMSs for EVs. Figure 2.26 lists some of these models. Their detailed description and corresponding mathematical equations can be found in the literature [3, 16]. Both the OFFPIM and the ONPIM can be applied to identify the model parameters in these battery models.

Table 2.12 Statistical results of voltage prediction error using the ONPIM at 10 °C.

Order	Maximum error (mV)	Average error (mV)	RMSE (mV)
$n = 0$	31.62	5.20	5.48
$n = 1$	28.51	2.15	2.62
$n = 2$	28.53	1.98	2.52
$n = 3$	27.72	1.91	2.45
$n = 4$	27.40	1.87	2.40
$n = 5$	26.30	1.84	2.36

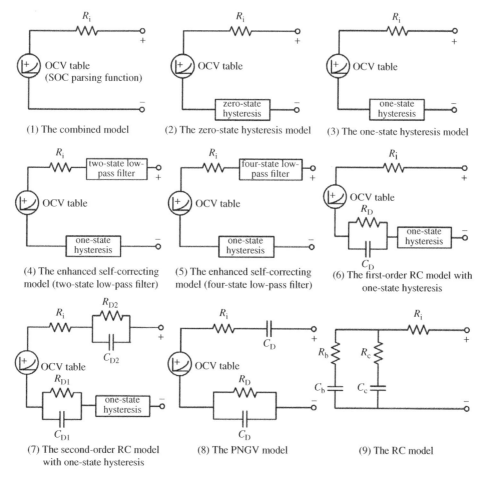

(1) The combined model

(2) The zero-state hysteresis model

(3) The one-state hysteresis model

(4) The enhanced self-correcting model (two-state low-pass filter)

(5) The enhanced self-correcting model (four-state low-pass filter)

(6) The first-order RC model with one-state hysteresis

(7) The second-order RC model with one-state hysteresis

(8) The PNGV model

(9) The RC model

Figure 2.26 Some battery models.

2.10 Summary

This chapter discusses battery modeling techniques which are the basis of states estimation for BMSs in EVs. In general, battery models can be grouped into three types: EMs, black box models, and ECMs.

EMs and black box models are introduced and ECMs with different orders are elaborated as they are more suitable for on-board use for BMSs in EVs. Based on ECMs, the OFFPIM and the ONPIM are explained in detail.

The applications of the OFFPIM and the ONPIM to six ECMs with different orders are explored by using the AIC indicating the balance between model accuracy and complexity. The results show that the first-order RC model, namely the Thevenin model, is the most suitable model structure for LiBs. Moreover, the influence of battery aging, battery type, and battery temperature on the OFFPIM and the ONPIM are discussed.

References

1 Schmidt, A.P., Bitzer, M., Imre, Á.W., and Guzzella, L. (2010). Experiment-driven electrochemical modeling and systematic parameterization for a lithium-ion battery cell. *Journal of Power Sources* 195: 5071–5080.

2 Kang, L., Zhao, X., and Ma, J. (2014). A new neural network model for the state-of-charge estimation in the battery degradation process. *Applied Energy* 121: 20–27.

3 Hu, X., Li, S., and Peng, H. (2012). A comparative study of equivalent circuit models for Li-ion batteries. *Journal of Power Sources* 198: 359–367.

4 Nejad, S., Gladwin, D.T., and Stone, D.A. (2016). A systematic review of lumped-parameter equivalent circuit models for real-time estimation of lithium-ion battery states. *Journal of Power Sources* 316: 183–196.

5 Fuller, T.F., Doyle, M., and Newman, J. (1994). Simulation and optimization of the dual lithium ion insertion cell. *Journal of the Electrochemical Society* 141: 1–10.

6 Shen, W.X., Chan, C.C., Lo, E.W.C., and Chau, K.T. (2002). A new battery available capacity indicator for electric vehicles using neural network. *Energy Conversion and Management* 43 (6): 817–826.

7 Eddahech, A., Briat, O., Bertrand, N. et al. (2012). Behavior and state-of-health monitoring of Li-ion batteries using impedance spectroscopy and recurrent neural networks. *International Journal of Electrical Power and Energy Systems* 42 (1): 487–494.

8 Klass, V., Behm, M., and Lindbergh, G. (2014). A support vector machine-based state-of-health estimation method for lithium-ion batteries under electric vehicle operation. *Journal of Power Sources* 270: 262–272.

9 Burgos, C., Sáez, D., Orchard, M.E., and Cárdenas, R. (2015). Fuzzy modelling for the state-of-charge estimation of lead-acid batteries. *Journal of Power Sources* 274: 355–366.

10 Chen, Z., Li, X., Shen, J. et al. (2016). A novel state of charge estimation algorithm for lithium-ion battery packs of electric vehicles. *Energies* 9: 710.

11 He, H., Zhang, X., Xiong, R. et al. (2012). Online model-based estimation of state-of-charge and open-circuit voltage of lithium-ion batteries in electric vehicles. *Energy* 39 (1): 310–318.

12 Xiong, R., Gong, X., Mi, C.C., and Sun, F. (2013). A robust state-of-charge estimator for multiple types of lithium-ion batteries using adaptive extended Kalman filter. *Journal of Power Sources* 243: 805–816.

13 Srinivasan, V. and Newman, J. (2006). Existence of path-dependence in the $LiFePO_4$ electrode. *Electrochemical and Solid-State Letters* 9 (3): A110–A114.

14 Sun, F., Xiong, R., and He, H. (2016). A systematic state-of-charge estimation framework for multi-cell battery pack in electric vehicles using bias correction technique. *Applied Energy* 162: 1399–1409.

15 Xiong, R., He, H., and Zhao, K. (2015). Research on an online identification algorithm for a Thevenin battery model by an experimental approach. *International Journal of Green Energy* 12: 272–278.

16 He, H., Xiong, R., and Fan, J. (2011). Evaluation of lithium-ion battery equivalent circuit models for state of charge estimation by an experimental approach. *Energies* 4: 582–598.

3

Battery State of Charge and State of Energy Estimation

3.1 Background

Many techniques have been developed to construct state of charge (SOC) estimators for battery management systems (BMSs) in electric vehicles (EVs) [1]. Due to the closed-loop estimation ability and strong inhibiting effect on noises, Kalman filter (KF)-based SOC estimators are studied extensively [2]. H infinity filters (HIFs) are also studied to estimate battery SOC owing to their robustness against disturbances in battery models and noises [3].

Nevertheless, the SOC indicates only capacity state rather than energy state on which the EV driving range is dependent. Therefore, battery state of energy (SOE) estimators, which provide the essential basis and security of energy deployment, have also been developed for BMSs [4] in EVs. This chapter mainly presents model-based SOC and SOE estimation methods using the extended Kalman filter (EKF) and HIF algorithms.

3.2 Classification

There are four major groups of battery SOC estimation methods, including a look-up-table-based method, an ampere-hour (Ah) integral method, data-driven estimation methods, and model-based estimation methods, as shown in Figure 3.1.

3.2.1 Look-Up-Table-Based Method

Battery SOC is strongly related to external (static) characteristic parameters, such as open circuit voltage (OCV). It can be obtained by using a look-up table to represent the relationship between the SOC and the measured external parameters [5]. Figure 3.2 shows the relationship of OCV versus SOC for a lithium-ion battery (LiB) cell. It can be seen that the OCV of the LiB cell increases monotonically as the SOC increases. Such a relationship can be used to infer the SOC through a measured OCV.

The relationship between the OCV and the SOC is widely used for the SOC estimation in BMSs for EVs. Such a relationship can also be used to effectively correct erroneous SOC estimations. However, it is difficult to obtain a precise OCV in real-time because the true OCV requires a battery to be rested for a long period of time, which is not practical for EV applications. Therefore, the look-up-table-based SOC estimation method is more suitable for implementation in a laboratory than in real EV applications.

Advanced Battery Management Technologies for Electric Vehicles, First Edition. Rui Xiong and Weixiang Shen.
© 2019 John Wiley & Sons Ltd. Published 2019 by John Wiley & Sons Ltd.

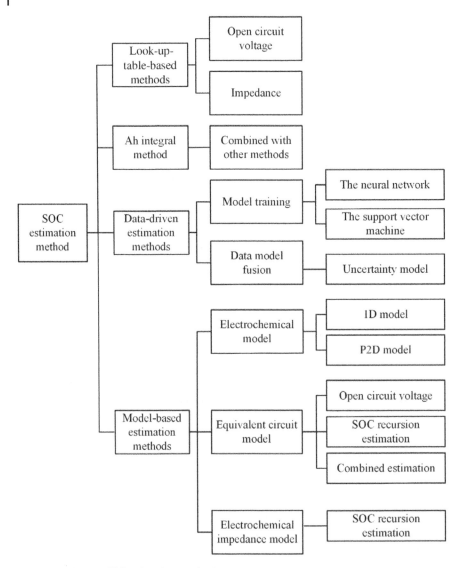

Figure 3.1 Battery SOC estimation methods.

3.2.2 Ampere-Hour Integral Method

The Ah integral method can calculate battery SOC with a known initial SOC as

$$z_k = z_0 - \int_{t_0}^{t_k} \eta_i i_L(t) dt / C_{max} \tag{3.1}$$

where z_k and z_0 denote the SOCs at the moment t_k and t_0, respectively, t_0 denotes the initial moment, $t_k = t_0 + k \times \Delta t$, Δt denotes the sampling interval, η_i denotes the coulomb efficiency for charge and discharge which can be determined by experiments, $I_L(t)$ denotes the load current on a battery, and C_{max} denotes the maximum available

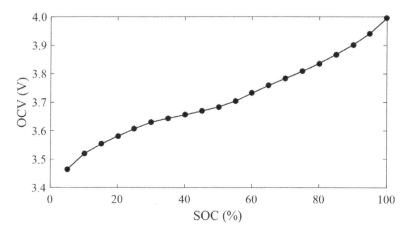

Figure 3.2 OCV versus SOC curve for a LiB cell.

capacity. It should be noted that C_{max} is generally defined as the nominal capacity which is the constant value given by manufacturers. The operating environment, aging conditions, and operating temperature will not affect the nominal capacity. However, they affect the maximum available capacity of a battery and thus EV driving distances. Therefore, C_{max} is defined as the maximum available capacity of a battery in this book, which is more suitable for EV applications.

The Ah integral method causes large SOC estimation errors owing to its three drawbacks. First, an accurate initial SOC is difficult to obtain in practice. Secondly, the measured current can be noisy which leads to accumulative errors by using Eq. (3.1). Thirdly, an accurate C_{max} value is hard to obtain in practice. Therefore, the Ah integral method generally will work with other techniques such as model-based methods to improve the SOC estimation accuracy.

3.2.3 Data-Driven Estimation Methods

Data-driven estimation methods use a large amount of experimental data of a system to develop a relationship between the input and output. The data-driven approach performs well in the following cases:

1. The global mathematical model of a system is completely unknown.
2. There are many uncertainties for a system model.
3. A system with uncertain structures in its operating process cannot be described by a mathematical model.
4. The mechanism model of a system is too complicated or its orders are too high, which becomes prohibitively difficult or impractical to analyze.

The data-driven estimation methods, also known as black box models, can be used to describe the strongly nonlinear relationship between the input (e.g. voltage and current) and output (e.g. SOC) of a battery. The application of black box models can realize parallel distributed processing and improve adaptive capability to deal with the complicated problem of the SOC estimation. The typical algorithms to construct black box models include fuzzy logical controller, neural network (NN), support vector machine and a combination of these algorithms. The NN is used to develop an SOC estimator in Shen [6], where the input layer contains the current, temperature, voltage of the battery, and the output layer only contains the SOC. The results show the NN can provide a high SOC estimation accuracy. However, the data-driven methods are very sensitive to the training data. If the training data has not covered the data generated in the current working conditions, the NN model established on the training data may provide the wrong SOC estimation.

3.2.4   Model-Based Estimation Methods

The model-based estimation methods use nonlinear state estimation algorithms and adaptive filters to estimate the SOC based on the state space equations derived from battery models. The typical algorithms include KFs [2], Luenberger observer [7], the proportion integration (PI) observer [8], the HIF [9] and the sliding-mode observer [10]. Figure 3.3 shows a general framework of model-based SOC estimation methods. The model-based estimation methods combine with the Ah integral method and the look-up table method describing the relationship between the OCV and SOC using a battery model. Their estimation process is explained as follows: an inaccurate SOC from the Ah integral method causes an erroneous OCV, which leads to the estimated terminal voltage errors. These errors are used to correct the SOC so that the estimated terminal voltage can track the measured terminal voltage. In other words, the OCV versus the SOC relationship can be used to obtain the correct SOC. Among all the algorithms, the KFs and HIFs have become the most general techniques for battery SOC estimation in EVs. These two methods will be given a detailed discussion in the following section.

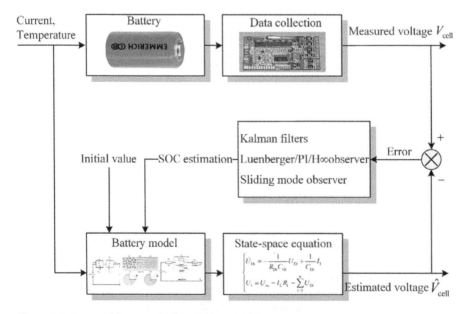

Figure 3.3 A general framework of model-based SOC estimation methods.

3.3 Model-Based SOC Estimation Method with Constant Model Parameters

Compared with the other three methods, the model-based method can estimate battery states in real-time during battery operation in EVs. Battery models (e.g. equivalent circuit models) are used to describe battery charge/discharge behaviors in the form of state space equations. The estimation process is closed-loop and robust against the modeling and process noises.

3.3.1 Discrete-Time Realization Algorithm

The optimal model structure determined in Chapter 2 is the Thevenin model, as shown in Figure 3.4. Its mathematical model is expressed as

$$\begin{cases} \dot{U}_D = -\dfrac{1}{C_D R_D} U_D + \dfrac{1}{C_D} i_L \\ U_t = U_{oc} - U_D - i_L R_i \end{cases} \tag{3.2}$$

According to the assumption in Chapter 2, the parameters of a battery cell are regarded as constant within the unit sampling time and the battery cell model can be linearized and regarded as a time varying system with constant parameters. Solving Eq. (3.2) in terms of the polarization voltage U_D gives

$$U_D(t) = \exp\left[-\frac{1}{C_D R_D}(t - t_0)\right] U_D(t_0) + \int_{t_0}^{t} \exp\left[-\frac{1}{C_D R_D}(t - t_0)\right] \frac{1}{C_D} i_L(\tau) d\tau \tag{3.3}$$

where t denotes the current time and t_0 denotes the initial time. τ is between $k\Delta t$ and $(k + 1)\Delta t$. By adopting $t_0 = k\Delta t, t = (k + 1)\Delta t, k = 0, 1, 2, \cdots$, Eq. (3.3) can be transformed into

$$U_D[(k + 1)\Delta t] = \exp\left(-\frac{\Delta t}{C_D R_D}\right) U_D(k\Delta t)$$

$$+ \int_{k\Delta t}^{(k+1)\Delta t} \exp\left\{-\frac{1}{C_D R_D}[(k + 1)\Delta t - \tau]\right\} \frac{1}{C_D} i_L(\tau) d\tau \tag{3.4}$$

Figure 3.4 One RC network – Thevenin model.

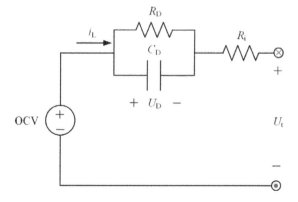

In unit sampling time, the measured current often has the sampling and holding characteristics, namely $u(\tau) = u[(k + 1)\Delta t]$, so the input $u[(k + 1)\Delta t]$ can be placed outside the integral. Equation (3.4) can be written as

$$U_D[(k + 1)\Delta t] = \exp\left(-\frac{\Delta t}{C_D R_D}\right) U_D(k\Delta t)$$

$$+ \int_{k\Delta t}^{(k+1)\Delta t} \exp\left\{-\frac{1}{C_D R_D}[(k + 1)\Delta t - \tau]\right\} d\tau \frac{1}{C_D} i_L[(k + 1)\Delta t] \tag{3.5}$$

Define $t = (k + 1)\Delta t - \tau$, $dt = -d\tau$. When the lower limit of the integral is $\tau = k\Delta t$, $t = (k + 1)\Delta t - k\Delta t = \Delta t$. When the upper limit of the integral is $\tau = (k + 1)\Delta t$, $t = (k + 1)\Delta t - \tau = 0$. Equation (3.5) can be further written as

$$U_D[(k + 1)\Delta t] = \exp\left(-\frac{\Delta t}{C_D R_D}\right) U_D(k\Delta t)$$

$$+ \int_0^{\Delta t} \exp\left(-\frac{1}{C_D R_D}\Delta t\right) d\tau \frac{1}{C_D} i_L[(k + 1)\Delta t] \tag{3.6}$$

After the integral of the second item on the right-side of Eq. (3.6), the polarization voltage can be calculated by

$$U_D[(k + 1)\Delta t] = \exp(-\Delta t/\tau)U_D(k\Delta t) + R_D i_L[(k + 1)\Delta t][1 - \exp(-\Delta t/\tau)] \tag{3.7}$$

where $\tau = R_D \times C_D$, if $U_{D,k+1}$ represents $U_D[(k + 1)\Delta t]$, $i_{L,k+1}$ represents $i_L[(k + 1)\Delta t]$, Eq. (3.7) can be simplified as

$$U_{D,k} = \exp(-\Delta t/\tau) \times U_{D,k-1} + [1 - \exp(-\Delta t/\tau)] \times i_{L,k} R_D \tag{3.8}$$

3.3.2 Extended Kalman Filter

The KF is an optimal estimation method for a linear system in the sense of minimum variance proposed by American scholar Kalman in the early 1960s. The EKF is an extension of the KF for its application to a nonlinear system. It provides a solution to handle random noises caused by the variation of battery model parameters and inaccurate measurement as variables in a state space equation. It makes full use of the measured data to estimate battery states by filtering the random noises of a battery system and its parameter measurement through a recursive method [2]. For any nonlinear time-varying system, it can be linearized as:

$$\begin{cases} \text{State equation}: \mathbf{x}_k = f(\mathbf{x}_{k-1}, \mathbf{u}_{k-1}) \approx \mathbf{A}_{k-1}\mathbf{x}_{k-1} + \mathbf{B}_{k-1}\mathbf{u}_{k-1} + \omega_{k-1} \\ \text{Observation equation}: \mathbf{y}_k = h(\mathbf{x}_k, \mathbf{u}_k) \approx \mathbf{C}_k\mathbf{x}_k + \mathbf{D}_k\mathbf{u}_k + v_k \end{cases} \tag{3.9}$$

where $f(\cdot)$ and $h(\cdot)$ are the state function and observation function of the nonlinear system, respectively; $\mathbf{x}$ is the n-dimension state vector; $\mathbf{u}$ is the r-dimension input vector; $\mathbf{y}$ is the m-dimension output vector; $\mathbf{A}$ is the system matrix of $n \times n$ dimension; $\mathbf{B}$ is the control matrix of $n \times r$ dimension; $\mathbf{C}$ is the output matrix of $m \times r$ dimension; $\mathbf{D}$ is the feedforward matrix of $m \times r$ dimension. w_{k-1} is the white noise of the system with zero mean, and its covariance is $\mathbf{Q}$; v_k is the white noise of the measurement with zero mean, its covariance is $\mathbf{R}$; w_{k-1} and v_k are independent of each other. The flowchart of the optimal estimation based on the EKF for battery states is summarized in Figure 3.5.

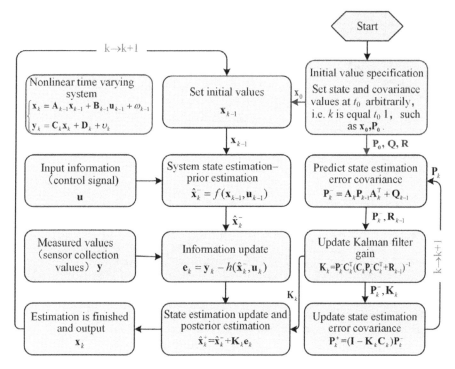

Figure 3.5 Flowchart of EKF-based estimation methods for battery SOC.

3.3.2.1 Selection of Correction Coefficients

The nature of the KF is an optimal recursive data processing algorithm. For most applications, the KF method is optimal and efficient. KF accomplishes the state or parameter estimation based on the monotonic relationship between the state of an observed system and observed values. The inaccurate state observations can bring observation errors; these errors are used to adjust the gain of the filter to realize the accurate estimation of the state or parameters. For a battery system represented by the Thevenin model, the four critical parameters are the OCV, dynamic characteristic parameters R_D and C_D, and ohmic internal resistance R_i. Figure 3.6 shows the model parameters for a LiB cell.

Figure 3.6 shows that only the OCV has a monotonic relationship with the SOC among these four critical parameters. A simple expression is used to represent such a monotonous relationship as

$$U_{oc}(z) = \alpha_0 + \alpha_1 z + \alpha_2 z^2 + \alpha_3 z^3 + \alpha_4/z + \alpha_5 \ln z + \alpha_6 \ln(1-z) \tag{3.10}$$

where $\alpha_i (i = 0, 1, \ldots, 5)$ are the fitting coefficients. This expression can enhance the natural correlation between the SOC and OCV and improve the convergence speed of the KF gain. Figure 3.7 shows the relationship between the change of the four parameters and the SOC.

3.3.2.2 SOC Estimation Based on EKF

The discretization of Eq. (3.1) leads to

$$z_k = z_{k-1} - \frac{\eta_i i_{L,k} \Delta t}{C_{max}} \tag{3.11}$$

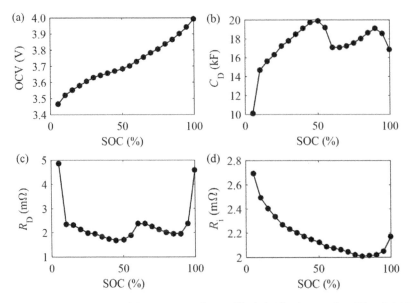

Figure 3.6 Thevenin model parameters obtained by hybrid pulse test for a LiB cell: (a) OCV; (b) polarization capacitance; (c) polarization internal resistance; (d) ohmic internal resistance.

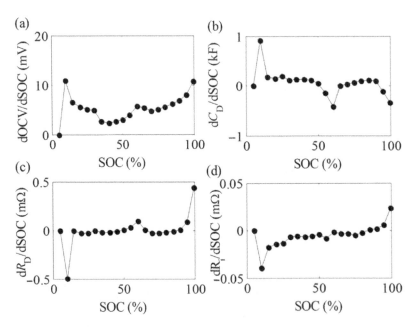

Figure 3.7 The change of Thevenin model parameters with SOC variation: (a) $dOCV/dSOC$; (b) $dC_D/dSOC$; (c) $dR_D/dSOC$; (d) $dR_i/dSOC$.

Based on Eqs. (3.10) and (3.11), the state-space equation of the battery can be obtained as

$$
\begin{cases}
\begin{bmatrix} U_{D,k} \\ z_k \end{bmatrix} = \begin{bmatrix} \exp\left(\dfrac{-\Delta t}{\tau}\right) & 0 \\ 0 & 1 \end{bmatrix} \begin{bmatrix} U_{D,k-1} \\ z_{k-1} \end{bmatrix} + \left\{ \begin{bmatrix} \left[1 - \exp\left(\dfrac{-\Delta t}{\tau}\right)\right] R_D \\ \dfrac{\eta_i \Delta t}{C_a} \end{bmatrix} \right\} \cdot i_{L,k} \\
U_{t,k} = \begin{bmatrix} -1 & \dfrac{\partial U_{oc}(z)}{\partial z} \end{bmatrix} \begin{bmatrix} U_{D,k-1} \\ z_{k-1} \end{bmatrix} + [-R_i] i_{L,k}
\end{cases}
\tag{3.12}
$$

where

$$
\frac{\partial U_{oc}(z)}{\partial z} = \alpha_1 + 2\alpha_2 z + 3\alpha_3 z^2 - \frac{\alpha_4}{z^2} + \frac{\alpha_5}{z} - \frac{\alpha_6}{1 - z}
\tag{3.13}
$$

After the establishment of the state space equation, the EKF can be used to estimate the SOC. Table 3.1 shows the procedure to calculate the SOC based on the EKF [2].

The offline parameter identification method (OFFPIM) in Chapter 2 can be used to determine the parameters of the battery model under different SOCs. Then, we can calculate the matrices **A**, **B**, **C**, and **D** at each time for the SOC estimation.

3.3.3 SOC Estimation Based on HIF

The KFs are the effective tools to estimate battery states based on the assumptions of accurate battery models and known statistical characteristics of noises [11]. Practically, such assumptions may not be valid for a battery operating in EVs. To achieve better estimation of the SOC, the HIF based on the minimax error criterion which was proposed and improved in [12, 13] can guarantee estimation accuracy in the worst case and it is increasingly popular in the field of battery state estimation. The following is a brief introduction to the HIF.

Consider a standard discrete-time linear system as:

$$
\begin{cases}
x_{k+1} = A_k x_k + w_{k-1} \\
y_k = C_k x_k + v_k
\end{cases}
\tag{3.14}
$$

The HIF is to minimize the largest estimation error as defined in the following cost function [12]:

$$
J_1 = \frac{\displaystyle\sum_{k=0}^{N-1} \|x_k - \hat{x}_k\|_{S_k}^2}{\|x_0 - \hat{x}_0\|_{P_0^{-1}}^2 + \displaystyle\sum_{k=0}^{N-1} (\|w_k\|_{Q_k^{-1}}^2 + \|v_k\|_{R_k^{-1}}^2)}
\tag{3.15}
$$

where $\hat{x}_k$ is guess values of x_k; P_0, Q_k, R_k, and S_k are the symmetric positive definite matrices that should be chosen by designers to suit the specific application. For example, we can make $S_k(1, 1)$ bigger than other elements of S_k if we are more interested in the estimation accuracy of the first element in x_k; in the same way, if we know that the second element of w_k is very small, we can make $Q_k(2, 2)$ smaller than the others. According

Table 3.1 Procedure for SOC calculation based on EKF.

The discretization equation of the Thevenin model can be expressed as

$$\begin{cases} \mathbf{x}_k = f(\mathbf{x}_{k-1}, \mathbf{u}_{k-1}) + \omega_{k-1} \approx \mathbf{A}_{k-1}\mathbf{x}_{k-1} + \mathbf{B}_{k-1}\mathbf{u}_{k-1} + \omega_{k-1} \\ \mathbf{y}_k = h(\mathbf{x}_k, \mathbf{u}_k) + v_k \approx \mathbf{C}_k\mathbf{x}_k + \mathbf{D}_k + v_k \end{cases}$$

where $\mathbf{x} = \begin{bmatrix} U_D z \end{bmatrix}^\mathrm{T}, \mathbf{u} = i_\mathrm{L}, \mathbf{y} = U_t;$

$$\mathbf{A}_k = \begin{bmatrix} \exp(-\Delta t/\tau) & 0 \\ 0 & 1 \end{bmatrix}, \mathbf{B}_k = \begin{pmatrix} (1 - \exp(-\Delta t/\tau))R_\mathrm{D} \\ \eta_i \Delta t/C_a \end{pmatrix},$$

$$\mathbf{C}_k = \begin{bmatrix} -1 & \dfrac{dU_{oc}(z)}{dz} \end{bmatrix}, \mathbf{D}_k = h(\mathbf{x}_k, \mathbf{u}_k) - \mathbf{C}_k\mathbf{x}_k = U_{oc,k} - U_{\mathrm{D},k} - R_\mathrm{i}\mathbf{u}_k - \mathbf{C}_k\mathbf{x}_k$$

(1) **Initialization**
Set the initial values of the state observer: $\mathbf{x}_0, \mathbf{P}_0, \mathbf{Q}, \mathbf{R}$

(2) **Prior-estimate prediction: Time update (state from time $(k-1)^+$ to time$(k)^-$)**
For $k = 1, 2, \ldots$, the following priori estimate (time update) operation is performed to deduce the state and covariance estimates from the previous time $(k-1)^+$ to the current time $(k)^-$. The EKF time update equation is expressed as follows:

System state estimation:

$$\widehat{\mathbf{x}}_k^- = f(\mathbf{x}_{k-1}, \mathbf{u}_{k-1})$$

Covariance estimation of error:

$$\mathbf{P}_k^- = \mathbf{A}_{k-1}\mathbf{P}_{k-1}\mathbf{A}_{k-1}^\mathrm{T} + \mathbf{Q}$$

(3) **Posteriori estimation-correction: measurement update (state from time $(k)^-$ to time $(k)^+$)**
Measured value Y_k at time k is used to correct the state and covariance estimation. The estimation results are expressed separately with $\widehat{\mathbf{x}}_k^+$ and $\mathbf{P}_k^+$. The EKF equation is updated as follows:

Information matrix:

$$\mathbf{e}_k = \mathbf{Y}_k - h(\widehat{\mathbf{x}}_k^-, \mathbf{u}_k)$$

Kalman gain matrix:

$$\mathbf{K}_k = \mathbf{P}_k^-\mathbf{C}_k^\mathrm{T}(\mathbf{C}_k\mathbf{P}_k^-\mathbf{C}_k^\mathrm{T} + \mathbf{R})^{-1}$$

System state correction:

$$\widehat{\mathbf{x}}_k^+ = \widehat{\mathbf{x}}_k^- + \mathbf{K}_k\mathbf{e}_k$$

Error covariance update:

$$\mathbf{P}_k^+ = (\mathbf{I} - \mathbf{K}_k\mathbf{C}_k)\mathbf{P}_k^-$$

Time scale update: State and covariance matrix are deduced from time $(k)^+$ to time $(k) = (k+1)^-$ for estimation at time $(k+1)$

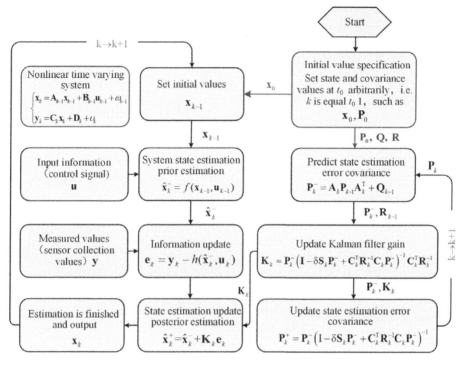

Figure 3.8 Flowchart of the HIF-based SOC estimation method.

to the mathematical deduction in [13], the following recursive filtering algorithm is obtained.

$$
\begin{cases}
K_k = P_k(I - \delta S_k P_k + C_k^T R_k^{-1} C_k P_k)^{-1} C_k^T R_k^{-1} \\
\hat{x}_{k+1} = A_k \hat{x}_k + A_k K_k (y_k - C_k \hat{x}_k) \\
P_{k+1} = A_k P_k (I - \delta S_k P_k + C_k^T R_k^{-1} C_k P_k)^{-1} A_k^T + Q_k
\end{cases}
\tag{3.16}
$$

Based on Eq. (3.16), the flowchart of the HIF based SOC estimation method is shown in Figure 3.8.

Based on the established state space equation of a battery system as shown in Eqs. (3.11)–(3.15), the procedure to calculate the SOC based on the HIF is shown in Table 3.2.

3.3.4 Case Study

The EKF and HIF are used to estimate the battery SOC. Figure 3.9 shows the estimation results under urban dynamometer driving schedule (UDDS) testing data at a temperature of 25 °C; the estimation statistics are listed in Table 3.3. The maximum error (MAE), standard deviation (STD), and mean error (MEE) are shown. Figure 3.9a,b demonstrates that the estimated terminal voltages based on the EKF and HIF are both close to the

Table 3.2 Procedure to calculate SOC based on HIF.

Initialization

Set the initial values of the state observer:

$$\mathbf{x}_0, \mathbf{P}_0, \mathbf{Q}, \mathbf{R}$$

δ and S_k are designed by users according to different requirements, $k \in \{1, 2, \ldots, \infty\}$

Step 1: Time-update equations

State estimate time update:

$$\widehat{\mathbf{x}}_k^- = f(\mathbf{x}_{k-1}, \mathbf{u}_{k-1})$$

Error covariance time update:

$$P_k^- = A_{k-1} P_{k-1}^+ A_{k-1}^T + Q_{k-1}$$

Step 2: Measurement-update equations

Gain matrix:

$$K_k = P_k^-(I - \delta S_k P_k^- + C_k^T R_k^{-1} C_k P_k^-)^{-1} C_k^T R_k^{-1}$$

State estimate measurement update:

$$\widehat{x}_k^+ = \widehat{x}_{k-1}^- + K_k(y_k - f(\widehat{x}_k^-, u_k))$$

Error covariance measurement update:

$$P_k^+ = P_k^-(I - \delta S_k P_k^- + C_k^T R_k^{-1} C_k P_k^-)^{-1}$$

Table 3.3 SOC estimation statistics under UDDS testing data at 25 °C.

State	Filter	MAE	STD	MEE
SOC (%)	EKF	0.71	0.28	−0.04
	HIF	1.24	0.28	−0.36
Voltage (mV)	EKF	127.7	13.0	−1.2
	HIF	178.7	12.8	−3.2

measured data. Table 3.3 shows the SOC estimation statistics. It can be seen that the MEEs of the estimated terminal voltages for both the EKF and the HIF are within 5 mV. The SOC estimation values based on the EKF and HIF agree with the reference SOC very well, and the MEEs of the estimated SOCs are within 1% as shown in Table 3.3.

3.3.5 Influence of Uncertainties on SOC Estimation

The uncertainties for a battery in EVs mainly include battery initial SOC value, dynamic working condition, and battery temperature. They will affect the SOC estimation accuracy. The following examples show the influence of these uncertainties on the SOC estimation.

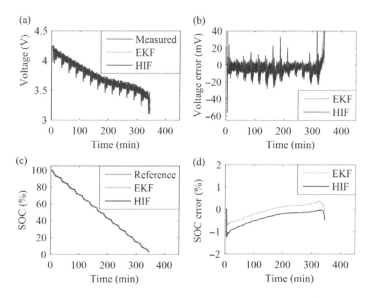

Figure 3.9 SOC estimation results under UDDS testing data at 25 °C: (a) voltage; (b) voltage error; (c) SOC; (d) SOC error.

3.3.5.1 Initial SOC Value

In practice, an initial SOC value is generally not available in EVs. The robustness of the EKF- and HIF-based SOC estimation methods against erroneous initial SOC values needs to be evaluated. Figures 3.10 and 3.11 show the SOC estimation results with erroneous initial SOCs using the two filter techniques. Figures 3.10 and 3.11, respectively, show the estimation results with the assumptions of 90% and 70% initial SOCs against the true initial SOC of 100%. Their estimation statistics are listed in Tables 3.4 and 3.5, respectively.

In Figure 3.10, the estimation results based on the two filters converge to the real data quickly, and after convergence the estimation results track the measured voltage accurately. Table 3.4 shows that the MEEs of the estimated terminal voltages using the two filters are still within 3 mV, whereas the MAEs of the estimated SOCs are both within 1.5% for the two filters after the convergence.

Figure 3.11 shows that after convergence, both the terminal voltage and SOC estimation results are close to the measured terminal voltage and reference SOC. The MEEs of the estimated terminal voltages based on EKF and HIF are −3.1 and −2.3 mV,

Table 3.4 SOC estimation statistics under UDDS testing data with an initial SOC of 90%.

State	Filter	MAE	STD	MEE
SOC (%)	EKF	0.97	0.30	−0.07
	HIF	1.15	0.28	−0.33
Voltage (mV)	EKF	128.3	12.1	−1.8
	HIF	177.1	13.0	−3.0

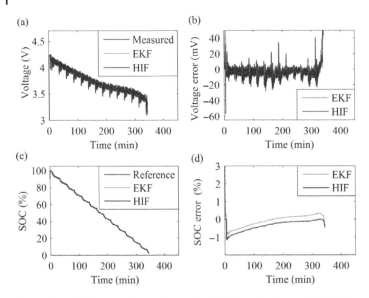

Figure 3.10 SOC estimation results under UDDS testing data at 25 °C with assumption of 90% erroneous initial SOC: (a) voltage; (b) voltage error; (c) SOC; (d) SOC error.

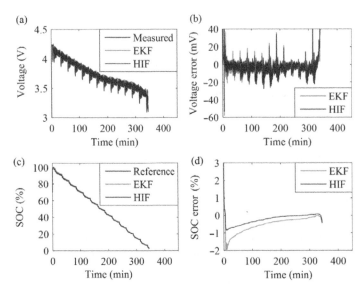

Figure 3.11 SOC estimation results under UDDS testing data at 25 °C with assumption of 70% erroneous initial SOC: (a) voltage; (b) voltage error; (c) SOC; (d) SOC error.

respectively. The MAEs of the estimated SOCs based on the EKF and HIF are 1.93 and 0.91%, respectively.

3.3.5.2 Dynamic Working Condition

Different EV driving cycles represent uncertainties of dynamic working condition. Figures 3.12 and 3.13 show the estimation results of the terminal voltage and SOC

Table 3.5 SOC estimation statistics under UDDS testing data with an initial SOC of 70%.

State	Filter	MAE	STD	MEE
SOC (%)	EKF	1.93	0.43	−0.51
	HIF	0.91	0.31	−0.20
Voltage (mV)	EKF	134.8	11.0	−3.1
	HIF	180.8	13.9	−2.3

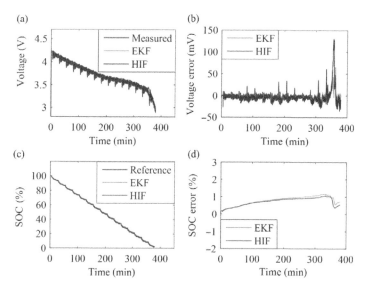

Figure 3.12 Estimation results of terminal voltage and SOC under UDDS testing data at 45 °C: (a) voltage; (b) voltage error; (c) SOC; (d) SOC error.

under two dynamic working conditions: UDDS and the new European driving cycle (NEDC), respectively; their corresponding estimation statistics are listed in Tables 3.6 and 3.7. Under UDDS, the terminal voltage estimations based on the two filters are both accurate, where the MAEs of the estimated terminal voltages are 91.3 mV for the EKF and 130.3 mV for the HIF as shown in Table 3.6, and the MAEs of the estimated SOCs for the EKF and HIF are both within 1.5%. Under NEDC, the terminal voltage estimations based on the two filters are both accurate too, where the MAEs of the estimated terminal voltages are 102.1 mV for the EKF and 152.2 mV for the HIF as shown in Table 3.7, and the MAEs of the estimated SOCs for both the EKF and HIF are within 1%. Therefore, the EKF- and HIF-based methods can provide the accurate estimation for battery terminal voltage and SOC for different dynamic working conditions.

3.3.5.3 Battery Temperature
Battery temperature is changing during EV driving. Figures 3.14 and 3.15 show the estimation results based on the EKF and HIF at temperatures of 25 and 0 °C, respectively; their corresponding estimation statistics are listed in Tables 3.8 and 3.9. It can

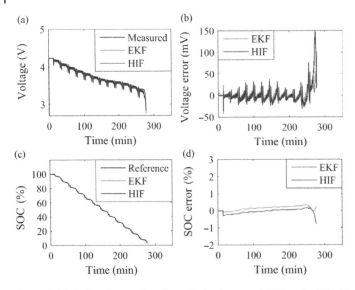

Figure 3.13 Estimation results of terminal voltage and SOC under NEDC testing data at 45 °C: (a) voltage; (b) voltage error; (c) SOC; (d) SOC error.

Table 3.6 SOC estimation statistics under UDDS testing data at 45 °C.

State	Filter	MAE	STD	MEE
SOC (%)	EKF	1.16	0.26	0.78
	HIF	1.06	0.23	0.72
Voltage (mV)	EKF	91.3	9.7	−2.4
	HIF	130.3	16.0	−2.5

Table 3.7 SOC estimation statistics under NEDC testing data at 45 °C.

State	Filter	MAE	STD	MEE
SOC (%)	EKF	0.33	0.11	0.14
	HIF	0.77	0.12	−0.04
Voltage (mV)	EKF	102.1	11.2	0.14
	HIF	152.2	22.3	1.36

be seen from Figure 3.14 and Table 3.8 that the voltage and SOC estimation results are accurate at the MAEs of the estimated terminal voltage and SOC within 125 mV and 1.5% for both filters. Figure 3.15 and Table 3.9 show that the terminal voltage and SOC estimation errors at 0 °C are larger than those at 25 °C as shown in Figure 3.14 at the MAEs of the estimated terminal voltage and SOCs within 165 mV and 5.0% for the EKF and 270 mV and 5.5% for the HIF, respectively. However, the terminal voltage and SOC

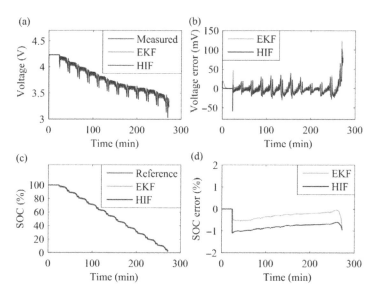

Figure 3.14 Estimation of terminal voltage and SOC under NEDC testing data at 25 °C: (a) voltage; (b) voltage error; (c) SOC; (d) SOC error.

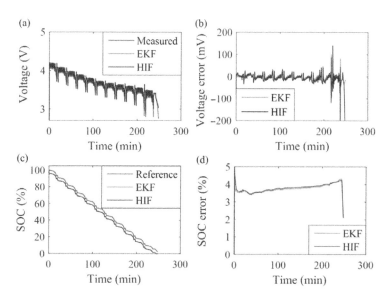

Figure 3.15 Estimation of terminal voltage and SOC under NEDC testing data at 0 °C: (a) voltage; (b) voltage error; (c) SOC; (d) SOC error.

estimation errors are mostly within 50 mV and 4% for both filters, indicating their high SOC estimation accuracy even at 0 °C.

Considering that the SOC estimation results are also accurate at 45 °C (Figures 3.12 and 3.13 and Tables 3.6 and 3.7), it can be concluded that both the EKF- and HIF-based methods can provide accurate SOC estimation at different temperatures.

Table 3.8 SOC estimation statistics under NEDC testing data at 25 °C.

State	Filter	MAE	STD	MEE
SOC (%)	EKF	0.88	0.16	−0.27
	HIF	1.10	0.25	−0.74
Voltage (mV)	EKF	123.0	12.9	−0.2
	HIF	102.7	12.5	−2.2

Table 3.9 SOC estimation statistics under NEDC testing data at 0 °C.

State	Filter	MAE	STD	MEE
SOC (%)	EKF	4.77	0.22	3.74
	HIF	5.46	0.23	3.79
Voltage (mV)	EKF	163.3	12.5	−4.7
	HIF	267.8	21.5	−6.9

3.4 Model-Based SOC Estimation Method with Identified Model Parameters in Real-Time

Battery performances vary with battery temperature, discharge/charge current and health state. To further improve the accuracy of SOC estimation, battery model parameters are identified by the recursive least squares (RLS) algorithm in real-time to reflect such variation. Then, the EKF algorithm is implemented on these battery models to provide more accurate SOC estimation.

3.4.1 Real-Time Modeling Process

Battery SOC estimation based on the real-time experimental data and the EKF is shown in Figure 3.16 [14]. It integrates the EKF for the SOC estimation as shown in Figure 3.5 and the RLS for the online parameter identification method discussed in Chapter 2. In this method, it is assumed that the maximum available capacity is known when calculating the response surface of the OCV.

The detailed procedure to calculate the SOC using model-based estimation method with identified model parameters in real-time is shown below.

Step 1: Real-time battery data acquisition. When the current or power excitation is loaded into the battery, the data acquisition module will collect the relevant characteristic data such as current, voltage, and temperature in real-time using the corresponding sensor. The software developed in the MATLAB/Simulink environment is used to run the acquisition module.

Step 2: Model parameter identification based on online measurement data. Based on the battery characteristic data measured in real-time, we can carry out the online identification of the battery model parameters using the RLS algorithm with

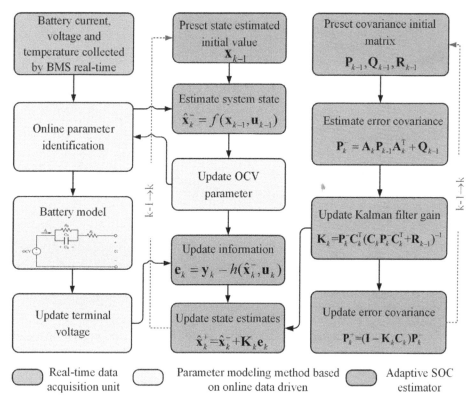

Figure 3.16 Battery SOC estimation method based on real-time experimental data and EKF.

a forgetting factor at each sampling interval, which allows the adaption of the battery model to the current operation conditions. First, the model parameters identified in real-time are used to estimate the system state and state estimation error covariance, i.e. calculate the prior state estimates and $\mathbf{P}_k^-$. Secondly, the OCV ($U_{oc}(\hat{z}_k^-)$) is adjusted by substituting the prior state estimate of the SOC into Eq. (3.10). Thirdly, the voltage prediction error e_k is updated based on the adjusted OCV. Finally, the voltage prediction error e_k is passed to the SOC estimator for generating a new message sequence.

Step 3: *SOC estimation based on EKF.* The system noises matrix R_{k-1} and the observed noises matrix Q_{k-1} are calculated by using the priori estimates $\mathbf{P}_k^-$ and the adaptive covariance matching algorithm at the time $k-1$. Then, an update operation of the KF gain matrix K_k is performed. Based on the model parameter update and the voltage prediction error, the posteriori estimation $\hat{\mathbf{x}}_k^+$, R_k, and Q_k are obtained via the error covariance and state update. Finally, the system state and its error covariance $\mathbf{P}_k^+$ are updated to complete the system state prediction at the time k and prepare the next time estimate.

The convergence of the developed model-based SOC estimation method in Figure 3.16 is ensured as follows. The online parameter identification method based on the RLS algorithm is able to update the error covariance matrix $\mathbf{P}$ and the data input matrix $\boldsymbol{\Phi}$ in real-time, which can adjust the least squares gain matrix K online to achieve accurate and reliable model parameters at the minimum sum of the mean

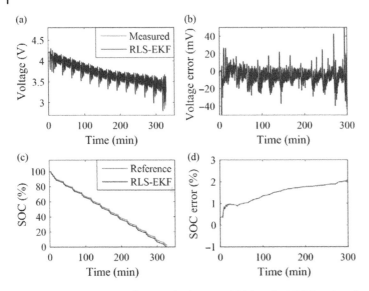

Figure 3.17 Estimation of terminal voltage and SOC under UDDS testing data at 0 °C: (a) voltage; (b) voltage error; (c) SOC; (d) SOC error.

squared errors. After obtaining accurate parameters, the SOC estimation method based on the EKF algorithm can reduce the prediction error of the battery terminal voltage by adjusting the SOC observation value. In this process, when the priori estimate of the battery SOC is determined, the OCV value of the battery can be updated in real-time by the OCV equation $U_{oc}(z)$. Only when the optimum OCV is determined can a minimal voltage prediction error be obtained. Therefore, the terminal voltage estimation error using the model-based technique is much smaller than that only using the online parameter identification technique.

3.4.2 Case Study

The model-based EKF method with identified model parameters in real-time is used to estimate the battery SOC. Figure 3.17 shows the estimation results for battery terminal voltage and SOC under UDDS testing data, where the initial value of the SOC is set to the correct value of 100%.

The estimation statistics are shown in Table 3.10. It can be seen that the MAE of the estimated terminal voltage is within 200 mV and the MAE of the estimated SOC is within 2.5%. These results show that the battery model with the parameters identified in real-time can accurately predict terminal voltage, and with this accurate battery model the EKF-based SOC estimation method can accurately track the reference SOC.

3.5 Model-Based SOE Estimation Method with Identified Model Parameters in Real-Time

The SOC calculation only considers charge flow into or out of a battery, which neglects energy losses caused by electrochemical reactions and internal resistances inside the

Table 3.10 SOC estimation statistics under UDDS testing data at 0 °C.

State	MAE	STD	MEE
SOC (%)	2.03	0.42	1.54
Voltage (mV)	188.3	10.8	−6.3

battery. These energy losses directly lead to changes of battery terminal voltage, which will decrease fast during discharging and increase slowly during charging. Terminal voltage will affect the battery energy state significantly. The SOE takes terminal voltage into the calculation which is able to indicate actual available energy when an EV is running. From an engineering point of view, the SOE is more practical than the SOC which is more meaningful for BMSs to predict the remaining driving range of pure EVs. In the following, the model-based method with identified model parameters in real-time will be introduced to estimate the SOE.

3.5.1 SOE Definition

The SOE reflects the residual energy of a battery, which is defined as the ratio of the remaining energy to the total available energy [15]. In this study, the SOE is expressed by [16]

$$z_k = z_{k-1} - \frac{\eta \Delta E_a}{E_a} = z_{k-1} - \frac{\eta_E U_{t,k-1} i_{L,k} \Delta t}{E_a} \tag{3.17}$$

where Δt represents the sampling time, ΔE_a represents the variation of battery energy during each sampling time, E_a represents the maximum available energy of a battery, and η_E denotes the energy efficiency of a battery.

3.5.2 State Space Modeling

Equation (3.17) is used to replace the SOC in Eq. (3.12) with the SOE; we obtain the state space equation to estimate the SOE

$$\begin{cases} \mathbf{x}_k = \mathbf{A}_{k-1}\mathbf{x}_{k-1} + \mathbf{B}_{k-1}\mathbf{u}_{k-1} + \omega_{k-1} \\ \mathbf{y}_k = \mathbf{C}_k\mathbf{x}_k + \mathbf{D}_k\mathbf{u}_k + \upsilon_k \end{cases} \tag{3.18}$$

where $\mathbf{x} = \begin{bmatrix} U_D & z \end{bmatrix}^T$, $\mathbf{u} = i_L$, $\mathbf{y} = U_t$, $\mathbf{A} = \begin{bmatrix} \exp{-\Delta t/\tau} & 0 \\ 0 & 1 \end{bmatrix}$, $\mathbf{B} = \begin{Bmatrix} [1 - \exp(-\Delta t/\tau)]R_D \\ \eta_i U_t \Delta t/E_a \end{Bmatrix}$, $C = \begin{bmatrix} -1 & \partial U_{oc}(z)/\partial z \end{bmatrix}$, $D = [-R_i]$.

The EKF-based method can be applied to estimate the SOE based on Eq. (3.18), where the battery model parameters are identified in real-time. This EKF method is the same as the battery SOC estimation method shown in Figure 3.16 except that battery terminal voltage needs to be taken into account. It integrates the RLS method to estimate the model parameters in real-time and the EKF-based method to estimate the SOE. During the filtering process, the state of estimation–open circuit voltage (SOE–OCV) curve is used to make the correction on the SOE estimation and obtain the accurate SOE.

3.5.3 Case Study

The Chinese Typical City Driving Cycle (CTCDC) as shown in Figure 3.18 is used as an example to show the model-based SOE estimation method with identified model parameters in real-time. Figure 3.18a shows the CTCDC profile of speed versus time. This CTCDC profile for a period of 1314 seconds is designed by analyzing the statistical data of a city electric bus in China. It is used to evaluate energy consumption and

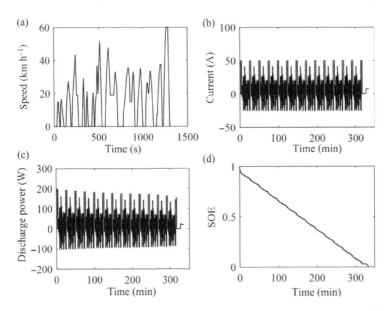

Figure 3.18 CTCDC profile: (a) speed versus time; (b) current versus time; (c) discharge power versus time; (d) SOE versus time.

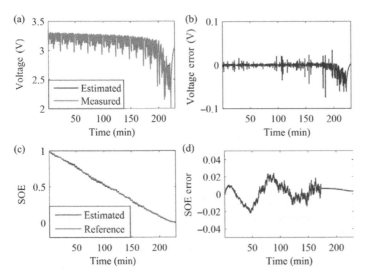

Figure 3.19 SOE estimation results under CTCDC at 0 °C: (a) voltage; (b) voltage error; (c) SOE; (d) SOE error.

Table 3.11 SOE estimation statistics under CTCDC testing data at 0 °C.

State	MAE	MEE
SOE (%)	2.39	0.22
Voltage (mV)	74.1	−2.6

economics of electric buses. To apply the CTCDC to test a battery, the conversion of the speed profile of the CTCDC into the current profile is required by running the simulation of an electric bus. Such a current profile is then loaded on a lithium iron phosphate (LFP) cell at 25 °C. Figure 3.18b–d shows the cycled profiles of the current, discharge power and the SOE versus time. The cell has the nominal voltage and capacity of 3.3 V and 27 Ah, respectively, with a lower cut-off voltage of 2 V and an upper cut-off voltage of 3.65 V.

Figure 3.19 shows the EKF-based SOE estimation results under the CTCDC testing data at 0 °C, and Table 3.11 shows the SOE estimation statistics. Figure 3.19a shows the measured and estimated terminal voltages and Figure 3.19b shows the voltage estimation error. It is observed that the estimated voltage agrees well with the measured voltage, and the MAE of the estimated terminal voltage is within 100 mV (Table 3.11). Figure 3.19c illustrates the estimated SOE against the reference SOE and Figure 3.19d shows the SOE estimation error. It can be seen that the SOE estimation results are almost the same as the reference SOE with an MAE of the estimated SOE of only 2.39%.

3.5.4 Influence of Uncertainties on SOE Estimation

Similar to the discussions of influence of uncertainties such as battery initial SOC value, dynamic working condition and temperature on the SOC estimation, the following examples show the influence of these uncertainties on the SOE estimation.

3.5.4.1 Initial SOE Value

Similar to the SOC, initial SOE values are generally not available in EVs. The robustness of the SOE estimator against erroneous initial SOE values is shown in Figure 3.20, where the initial SOEs are set to the erroneous values of 80% and 90% against the reference SOE of 100%. The SOE estimation statistics are listed in Table 3.12. Figure 3.20a,b shows the measured and estimated terminal voltages as well as the voltage estimation error. Figure 3.20c,d shows the estimated and reference SOE as well as the SOE estimation error.

It is observed that the estimated terminal voltage converges to the measured terminal voltage very quickly after a few seconds. This is also reflected in the SOE estimation which converges to the reference SOE quickly. The MAEs of the estimated terminal voltages are within 51.2 mV for 80% and 90% initial SOEs and the MAEs of the estimated SOEs are both mainly within 2% after the SOE estimation reaches steady state. The reason for obtaining these accurate estimation results is as follows. When the EKF-based SOE estimation method receives the erroneous initial SOE, a larger error of the terminal voltage is generated between the estimated and measured values. This larger error can be

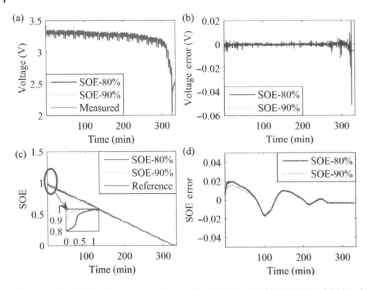

Figure 3.20 SOE estimation results under CTCDC at 25 °C with initial SOE values of 80% and 90%: (a) voltage; (b) voltage error; (c) SOE; (d) SOE error.

Table 3.12 SOE estimation statistics under CTCDC testing data with an initial SOE of 80% and 90%.

Initial SOE	State	MAE	MEE
90%	SOE (%)	1.73	0.1252
	Voltage (mV)	51.2	−0.1281
80%	SOE (%)	2.03	0.2254
	Voltage (mV)	51.2	−0.1260

used to adjust the KF gain to compensate the SOE estimation in an efficient closed-loop feedback.

3.5.4.2 Dynamic Working Condition

To indicate the ability of the model-based SOE estimation against dynamic working conditions, another working condition under NEDC is used. The SOE estimation results are presented in Figure 3.21. Figure 3.21a,b shows the estimated and measured terminal voltages as well as the voltage estimation error. Figure 3.21c,d shows the estimated and reference SOE values as well as their SOE estimation error. Table 3.13 shows that the MAE of the estimated terminal voltage is 55.6 mV and the MAE of the estimated SOE is 2.14%.

3.5.4.3 Battery Temperature

Figure 3.22 shows the model-based SOE estimation results at 25 °C, and the corresponding estimation statistics are listed in Table 3.14. The model-based method can estimate the terminal voltage and SOE accurately; the MAE of the estimated terminal voltage and

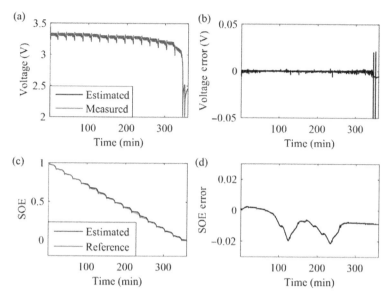

Figure 3.21 SOE estimation results under NEDC testing data at 25 °C: (a) voltage; (b) voltage error; (c) SOE; (d) SOE error.

Table 3.13 SOE estimation statistics under NEDC testing data at 25 °C.

State	MAE	MEE
SOE (%)	2.14	−0.80
Voltage (mV)	55.6	−0.18

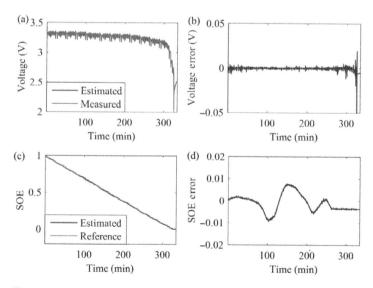

Figure 3.22 SOE estimation results under CTCDC at 25 °C: (a) voltage; (b) voltage error; (c) SOE; (d) SOE error.

Table 3.14 SOE estimation statistics under CTCDC testing data at 25 °C.

State	MAE	MEE
SOE (%)	0.93	−0.09
Voltage (mV)	51.2	−0.13

SOE are 51.2 mV and 0.93%, respectively. Based on the SOE estimation results at 0 °C shown in Figure 3.19, it can be concluded that the model-based method can provide accurate SOE estimation at different temperatures.

3.6 Summary

This chapter presents four groups of battery SOC estimation methods. They include look-up-table-based methods, an Ah integral method, model-based estimation methods, and data-driven estimation methods.

Among these four groups of methods, the model-based estimation methods have been discussed in detail. They can be further divided into the model-based estimation method with constant model parameters and the model-based estimation method with identified model parameters in real-time. The former includes the EKF- and HIF-based methods to estimate the SOC against the uncertainties of initial SOC values, dynamic working conditions and temperatures. The latter only includes the EKF-based method to estimate the SOC.

The model-based estimation method with identified model parameters in real-time is also used to estimate the SOE against the uncertainties of battery initial SOE values, dynamic working conditions, and temperature.

References

1 Lu, L., Han, X., Li, J. et al. (2013). A review on the key issues for lithium-ion battery management in electric vehicles. *Journal of Power Sources* 226: 272–288.

2 Plett, G.L. (2004). Extended Kalman filtering for battery management systems of LiPB-based HEV battery packs: part 3. State and parameter estimation. *Journal of Power Sources* 134 (2): 277–292.

3 Zhang, X.M. and Han, Q.L. (2015). Event-based H∞ filtering for sampled-data systems. *Automatica* 51: 55–69.

4 Zhang, Y., Xiong, R., He, H., and Shen, W. (2017). Lithium-ion battery pack state of charge and state of energy estimation algorithms using a hardware-in-the-loop validation. *IEEE Transactions on Power Electronics* 32 (6): 4421–4431.

5 Xiong, R., Tian, J., Mu, H., and Wang, C. (2017). A systematic model-based degradation behavior recognition and health monitoring method for lithium-ion batteries. *Applied Energy* 207: 372–383.

6 Shen, Y. (2010). Adaptive online state-of-charge determination based on neuro-controller and neural network. *Energy Conversion and Management* 51 (5): 1093–1098.

7 Hu, X., Sun, F., and Zou, Y. (2010). Estimation of state of charge of a lithium-ion battery pack for electric vehicles using an adaptive Luenberger observer. *Energies* 3 (9): 1586–1603.

8 Xu, J., Mi, C.C., Cao, B. et al. (2014). The state of charge estimation of lithium-ion batteries based on a proportional-integral observer. *IEEE Transactions on Vehicular Technology* 63 (4): 1614–1621.

9 Zhang, F., Liu, G., Fang, L., and Wang, H. (2012). Estimation of battery state of charge with H-infinity observer: applied to a robot for inspecting power transmission lines. *IEEE Transactions on Industrial Electronics* 59 (2): 1086–1095.

10 Kim, I.S. (2006). The novel state of charge estimation method for lithium battery using sliding mode observer. *Journal of Power Sources* 163 (1): 584–590.

11 Xiong, R., Gong, X., Mi, C.C., and Sun, F. (2013). A robust state-of-charge estimator for multiple types of lithium-ion batteries using adaptive extended KF. *Journal of Power Sources* 243: 805–816.

12 Banavar, R.N. (1992). *A Game Theoretic Approach to Linear Dynamic Estimation*. Austin: Texas University.

13 Shen, X. (1995). Discrete H∞ filter design with application to speech enhancement. *IEEE International Conference on Acoustics, Speech and Signal Processing* 2: 1504–1507.

14 Smith, K.A., Rahn, C.D., and Wang, C.Y. (2007). Control oriented 1D electrochemical model of lithium ion battery. *Energy Conversion and Management* 48 (9): 2565–2578.

15 Mamadou, K., Lemaire, E., Delaille, A. et al. (2012). Definition of a state-of-energy indicator (SoE) for electrochemical storage devices: application for energetic availability forecasting. *Journal of the Electrochemical Society* 159 (8): A1298–A1307.

16 Wang, Y., Zhang, C., and Chen, Z. (2014). A method for joint estimation of state-of-charge and available energy of LiFePO$_4$ batteries. *Applied Energy* 135: 81–87.

4

Battery State of Health Estimation

4.1 Background

The capability of a battery to store energy and provide a certain power decreases over battery lifetime because of aging. The state of health (SOH) is an indicator to quantify aging level of a battery in terms of capacity fade and/or power fade. It can be used to recognize an ongoing degradation of a battery, prevent its possible failure, and decide the replacement and removal of a battery system (or a battery pack) from electric vehicles (EVs).

The battery SOH can be defined as battery capacity decrease and internal resistance (or impedance) increase. It reflects the health of a battery cell in percentage; 100% stands for a fresh cell. When the battery capacity drops to 80% of the initial rated capacity, namely the SOH is 80%, the battery should be considered for replacement in an EV. Similarly, when the battery internal resistance doubles, which results in a significant power drop, the battery should also be considered for removal from an EV [1].

The SOH can be generally estimated by experimental and model-based methods. The experimental methods analyze the stored data which affect battery lifetime during its cyclic operation to directly determine the SOH. The model-based methods determine the SOH by estimating battery parameters that are senstive to degradation of a battery cell throughout the real-time operation of the battery. Each group of methods can be further divided into different methods, as summarized in Figure 4.1. In the following, we will consider all of these methods and their strengths and weaknesses for use in online battery management system (BMS) applications and discuss a potential and promising method to estimate the SOH in EV applications.

4.2 Experimental Methods

The experimental methods monitor battery behaviors by analyzing the full cycle of experimental data of battery voltage, current, and temperature. They can be further divided into direct measurement methods and indirect analysis methods.

Advanced Battery Management Technologies for Electric Vehicles, First Edition. Rui Xiong and Weixiang Shen.

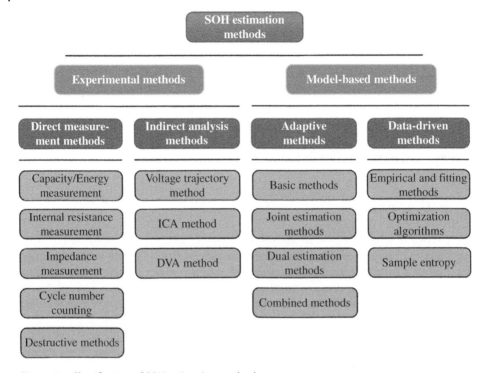

Figure 4.1 Classification of SOH estimation methods.

4.2.1   Direct Measurement Methods

The direct measurement methods monitor the directly measurable parameters related to battery lifetime, such as battery capacity or energy, ohmic resistance, impedance, and cycle number counting.

4.2.1.1   Capacity or Energy Measurement

The SOH can be determined by capacity/energy of a battery. Ampere-hour (Ah) counting is commonly used to directly calculate the capacity/energy of a battery through the high precision measurement of current and/or voltage. The SOH is the ratio of the capacity or energy currently available to that of a fresh battery [2]. This method is only suitable for laboratory use. In order to solve the problem that battery capacity/energy cannot be measured in real-time, a number of online estimation methods and combination methods using both offline measurement and online estimation have been proposed with the development of the BMS.

4.2.1.2   Internal Resistance Measurement

When the internal resistance of a battery increases to twice that of a fresh battery, the battery should be removed in EVs. For lithium-ion batteries (LiBs), the internal resistance includes ohmic resistance and polarization resistance. Ohmic resistance is determined by electrode material, electrolyte, separator resistance and contact resistance. In normal working conditions, the ohmic resistance contributes mostly to voltage drop,

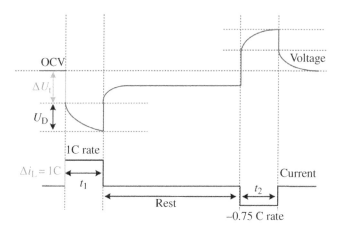

Figure 4.2 Current and voltage profile in a discharge and charge pulse.

which is basically linear to current density. Figure 4.2 plots a current and voltage profile in a discharge and charge pulse.

Similar pulses can be obtained when the EV is in the operation condition of braking or accelerating, or in the pulse test explained in Chapter 2. According to Ohm's law, the ohmic resistance can be calculated by

$$R_i = \frac{\Delta U_t}{\Delta i_L} \tag{4.1}$$

where Δi_L stands for the pulse current and ΔU_t denotes the pulse voltage. It also shows an additional voltage drop which determines polarization voltage U_D. With the internal resistance, the SOH can also be calculated. Figure 4.2 shows an example of the pulse. It is noted that the current and voltage of the pulse can be taken as other values. This ohmic resistance can be identified based on equivalent circuit models (ECMs) in real-time with the model-based method introduced in Section 2.6.

4.2.1.3 Impedance Measurement

Electrochemical impedance spectroscopy (EIS) is one of the most promising methods to characterize aging effects for LiBs. The impedance measured through EIS exhibits inductive effects due to battery wiring and porous structure at high frequency, capacitive effects due to diffusion at low frequency, and purely ohmic effects at the range of intermediate frequencies. The impedance of the middle frequency range increases with the battery aging and indicates the SOH [3], as shown in Figure 4.3. The EIS provides information on different aging mechanisms and can be used as a diagnostic tool.

4.2.1.4 Cycle Number Counting

The number of cycles or accumulated charge in the whole life cycle of a battery can be counted as the basis of the life model. It is commonly used to indicate battery SOH in a laptop or other small electronic products. In general, one cycle can be recorded after a battery is fully charged with the constant current constant voltage (CC/CV) charging method at the constant charging current of 1C and fully discharged with the constant discharging current of 1C. If the total life cycle number of a battery is N_{total} (normally

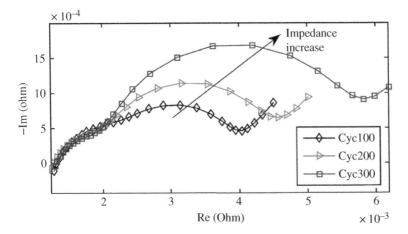

Figure 4.3 EIS results under different aging status.

given by manufacturers) and the current cycle number the battery has experienced is N_{exp}, then the battery SOH can be calculated as $(N_{total} - N_{exp})/N_{total}$.

4.2.1.5 Destructive Methods

These methods need to destroy a cell. After the cell is disassembled, scanning electron microscopy (SEM), scanning transmission electron microscopy, X-ray photoelectron spectroscopy, Raman spectroscopy, X-ray diffraction, and cyclic voltammetry are widely used to examine the cell. As shown in Figure 4.4, the SEM method is used to obtain the images of the internal contents of the cell. It can be observed that as the cycle number increases, stress is produced and accumulated in the active particles which can cause cracks spreading at the grain boundary of the particles, leading to mechanical failure of the anode materials. Generally, the examination offers the exact degradation information which can indicate the SOH. However, these methods can only be conducted in a laboratory.

4.2.2 Indirect Analysis Methods

With the entire degradation data of the major parameters affecting battery life, indirect analysis methods are proposed to calculate the SOH. They can be further divided into adaptive methods and data-driven methods. Generally, indirect analysis methods are multi-step derivation methods using health parameters strongly related to degradation of battery capacity or internal resistance. After the relationship between the health parameters and the SOH is established, the health parameters can be used to estimate the SOH. The health parameters include voltage trajectory or charging time of CC/CV at constant voltage (CV) step, incremental capacity analysis (ICA), and differential voltage analysis (DVA).

4.2.2.1 Voltage Trajectory Method

Battery terminal voltage can be used to characterize battery SOH. The CC/CV method is widely employed to charge batteries. Figure 4.5 shows an example of the charging profile

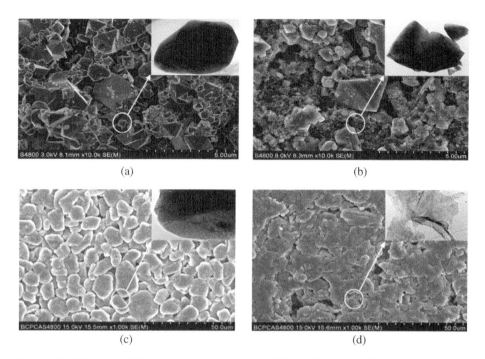

(a)

(b)

(c)

(d)

Figure 4.4 SEM images of fresh and aged battery cells: (a) fresh cell cathode; (b) aged cell cathode; (c) fresh cell anode; (d) aged cell anode.

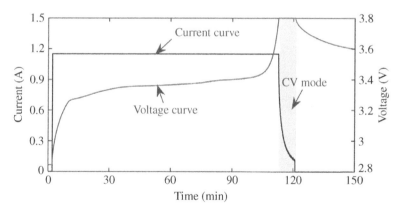

Figure 4.5 Current and voltage profiles of the CC/CV charging method.

for an A123 ANR26650 lithium iron phosphate (LFP) cell, where the nominal capacity of the battery is 2.3 Ah. The charging method consists of two steps: the constant current (CC) step where the CC is applied to charge a battery until the upper cut-off voltage of the battery is reached; and then the CV step where the battery remains in floating mode at the upper cut-off voltage until the current reaches the minimum threshold, generally in the range of C/10–C/30.

The CC/CV curve from 0 to 100% of the state of charge (SOC) will change with the battery degradation process. For each battery with the same chemistry, the total

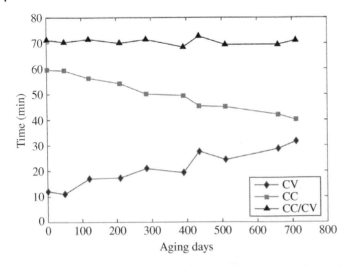

Figure 4.6 Charging times for CC, CV and CC/CV for a NMC LiB from Kokam.

charging time remains the same; the charging time at the CV step gradually increases and the charging time at the CC step gradually decreases with the degree of aging. From Figure 4.5, it is observed that the charging curve at the CC step gradually shrinks to the left and the voltage "platform" at the CV step gradually expands to the right. This charging time variation can be used to calculate the SOH. For four types of batteries with the CC/CV method, the SOH is calculated [4]. It is found that the charging time for the CC step decreases with aging level for all these batteries while the total charging time is constant. Figure 4.6 plots the total charging time for the CC/CV charging method at 1C, the charging time for the CC step, and the charging time for the CV step at different aging days. Through all the aging data collected during the operations, the relationship between the ratios of the charging time at the CC step to that at the CV step and aging levels can be established to calculate the SOH.

4.2.2.2 ICA Method

To better understand battery degradation during the cell aging process through measurement of voltage and current in different duty regimes, an ICA method is used to derive time-correlated degradation behavior under cyclic conditions [5]. The integrated circuit (IC) curves of a capacity change associated with a successive voltage step (dQ/dV) in the charge or discharge are obtained. Each of the peaks in the IC curve, representing an electrochemical process that takes place inside a cell, has a unique shape, intensity, and position. Any change in position and shape of an IC peak is a symptom of cell aging. Figure 4.7 shows the voltage–capacity curves and related incremental capacity signatures of charge and discharge regimes of C/20 and C/3 at 25 °C for a 90 Ah $LiMn_2O_4$ cell. The peaks observed in the IC curves correspond to the stage in the graphite negative electrode (NE), convoluting with phase transformation on the positive electrode (PE) [6].

In the graphite NE, lithium intercalation transforms C to LiC_6 in at least five distinct stages, noted as ⑤–①. In the PE, two phase transformations take place in $LiMn_2O_4$ at 4.1 and 4.0 V, respectively, denoted as I and II. Due to the distinct stages in the NE,

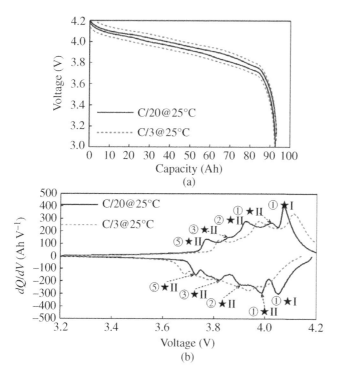

Figure 4.7 (a) Voltage–capacity curves and (b) IC of the charge and discharge regimes.

the IC peak corresponding to reaction II in the PE is split into three peaks, as shown in the IC curve (black solid line) in Figure 4.7b. The same phenomenon happens in the NE. Thus, the peak marked as ①★I represents reaction ① in the NE convoluting with reaction I in the PE. The IC curves show greater sensitivity than the conventional charge and discharge curves that allow us to probe any gradual changes of cell behavior during the thermal aging process.

In order to track the SOH and understand aging mechanisms, IC curves are always excited by a very low current, such as C/25. However, it is very hard to accurately capture smaller current in practical applications. A large current, such as C/3, has also been used to analyze IC behaviors. Figure 4.7b shows that there are only three apparent peaks in the IC curve of charge regime at C/3, denoted as ①★I, ②★II, and ⑤★II, respectively. The area under each peak represents the whole incremental capacity involved in the associated reaction. It indicates that the IC curve with the large current can also provide degradation information.

Figure 4.8 shows the comparison of IC curves for cells cycling at 40 °C and 10 °C with a depth of discharge (DOD) of 50% in different thermal cycling conditions, which are derived from the charge regimes in the capacity test using C/3 rate at 25 °C, as a function of the cycle number. It is observed that the IC peak is a useful indicator for connecting battery degradation to voltage behavior. As the cycle number increases, the peaks move down to the right. The battery SOH can be deduced by the relationship between cycle number and the peaks.

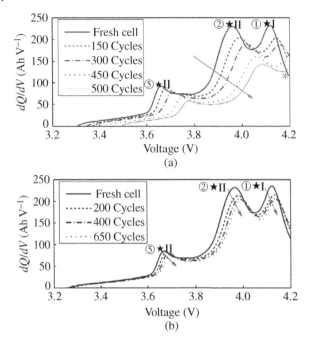

Figure 4.8 Evolution of IC curves for cells cycling at DOD of 50% at (a) 40 °C, and (b) 10 °C.

4.2.2.3 DVA Method

The DVA method can also be used to analyze battery degradation. One unique advantage of the DVA method is that clearly identifiable peaks appear in the DVA curve of a LiB. Although IC curves can help to understand cell degradation, the DVA curves are shown to be capable of identifying many mechanisms contributing to battery degradation. The mechanisms include loss of cathode active material, loss of anode active material, and loss of lithium inventory. Figure 4.9 shows the voltage–capacity curve and the DVA curve at C/3 and a temperature of 25 °C.

In the DVA curve, the staging phenomena become valleys while the peaks represent non-stoichiometry in the single-phase regions (solid solution). The DVA curve is naturally divided into three regions by the peaks, where each region corresponds to the capacity of a phase transformation involved in the reaction. The charging capacities corresponding to each region, denoted as Q_A, Q_B, and Q_C, are associated with the areas under the peaks ⑤★II, ②★II, and ①★I, respectively, in the IC curve. Thus, the aging mechanism of the cell can be similarly analyzed by investigating the changes in these phases. The same conclusion can be obtained from the DVA curves as the ICA curves. The comparison of the DVA curves for the cells cycling at 40 °C and 10 °C with a DOD of 50% at different cycle numbers is shown in Figure 4.10.

Figure 4.11 shows the evolution of Q_A, Q_B, and Q_C for the cells cycling to a DOD of 50% at 40 °C and 10 °C in different thermal cycling conditions, which are derived from the DVA curves as a function of the cycle number. It can be seen that Q_A, Q_B, and Q_C decrease significantly as the cycle number increases. Q_A and Q_B decrease at nearly

Figure 4.9 Voltage–capacity curve and DVA curve at CC charging regime at C/3 and 25 °C.

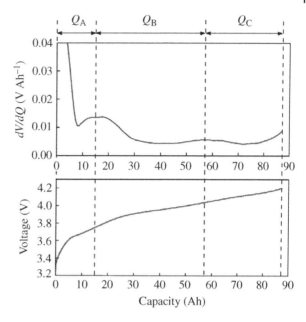

Figure 4.10 Evolution of DVA curves for cells cycling at DOD of 50% at (a) 40 °C, and (b) 10 °C.

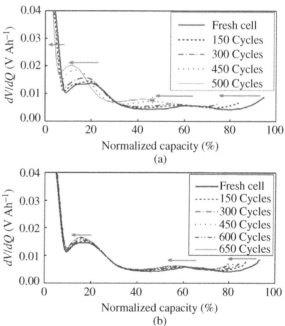

the same rate, indicating that the capacity loss arises from the loss of active material in the NE. Q_C decreases at a higher rate, indicating that a loss of lithium inventory exists in addition to the loss of active material. Thus, the mapping between Q_B and battery degradation is suitable for battery SOH estimation.

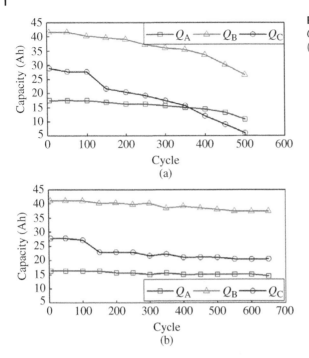

Figure 4.11 Evolution of Q_A, Q_B, and Q_C for cells cycling at DOD of 50% at (a) 40 °C, and (b) 10 °C.

4.3 Model-Based Methods

Model-based estimation methods can be further divided into adaptive state estimation methods and data-driven methods. The adaptive state estimation methods consider capacity, resistance and other parameters as slow change states of a battery model (e.g. ECM), where the estimation process involves the closed-loop control and feedback. These parameters are estimated to calculate the SOH. The data-driven methods model the historic data of the battery parameters affecting battery life by using black box models, machine learning, and intelligent optimization algorithms.

4.3.1 Adaptive State Estimation Methods

Adaptive state estimation methods are very promising in the use of online BMSs in EVs. In Chapter 2, we introduced the flowcharts of the offline and online parameter identification methods to estimate parameters of a battery model. With the estimated resistance, impedance, open circuit voltage (OCV) and other characterization parameters, the battery SOH can be obtained. Figure 4.12 shows a general operation framework to identify battery model parameters through adaptive filtering or observer algorithms.

A Kalman filter (KF) is commonly employed to estimate battery model parameters and states. The detailed computation procedure can be found in Chapter 3. There are many derived and improved algorithms based on the KF, such as extended Kalman filter (EKF), unscented Kalman filter (UKF), particle filter (PF), robust extended Kalman filter (REKF), adaptive extended Kalman filter (AEKF), adaptive unscented Kalman filter (AUKF), and adaptive particle filter (APF). The recursive least squares (RLS) method

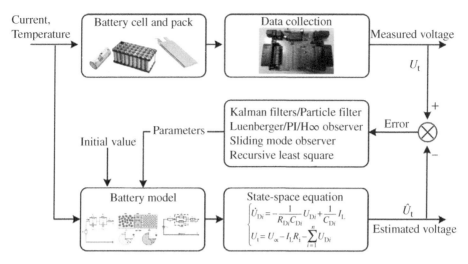

Figure 4.12 General operation framework to identify battery model parameters.

is also a filtering algorithm and its detailed application in parameter identification can be found in Section 2.6. In addition, nonlinear state observers, such as the Luenberger observer, proportional integral (PI) observer, H infinite observer, and sliding mode observer, are also used to estimate battery parameters and states.

The AEKF algorithm is used to identify the parameters of the Thevenin model [7]. Its implementation procedure is the same as the SOC estimation presented in Figure 3.5. There are two steps to implement an AEKF algorithm. In the first step, a prediction state is required, where the filter estimates the current output variable. In the second step, the estimation is updated in order to obtain a more accurate result and give the estimation a higher certainty. It requires the use of recursive equations based on the state space equation derived from a discrete battery model. The AEKFs for parameter identification [7] and state estimation [8] have different state matrices.

For Thevenin model presented in Section 2.4.2.2, a state space equation can be expressed by

$$
\begin{cases}
\dot{U}_{\mathrm{D}} = -\dfrac{1}{C_{\mathrm{D}}R_{\mathrm{D}}}U_{\mathrm{D}} + \dfrac{1}{C_{\mathrm{D}}}i_{\mathrm{L}} \\
U_{\mathrm{t}} = U_{\mathrm{oc}} - U_{\mathrm{D}} - i_{\mathrm{L}}R_{\mathrm{i}}
\end{cases}
\tag{4.2}
$$

where R_{i} is the ohmic resistance, R_{D} and C_{D} are polarization resistance and polarization capacity, respectively, i_{L} is the load current, U_{t} is the terminal voltage, U_{D} is the polarization voltage, and U_{oc} is the OCV.

We can define a state matrix $\mathbf{X}$ in the parameter identification as

$$
\mathbf{X} = \begin{bmatrix} U_{\mathrm{oc}} & U_{\mathrm{t}} & U_{\mathrm{D}} & 1/C_{\mathrm{D}} & 1/R_{\mathrm{D}} & R_{\mathrm{i}} \end{bmatrix}^{T}
\tag{4.3}
$$

Then, the state transition function can be obtained by

$$
\dot{\mathbf{X}} = f(\mathbf{X}, \mathbf{u}) = \begin{bmatrix} f_1 & f_2 & f_3 & f_4 & f_5 & f_6 \end{bmatrix}^{T}
\tag{4.4}
$$

where f_i denote

$$\begin{cases} f_1 = 0 \\ f_2 = x_1x_4x_5 - x_2x_4x_5 - (x_4x_5x_6 + x_4)u - x_6\dot{u} \\ f_3 = x_4u - x_3x_4x_5 \\ f_4 = 0 \\ f_5 = 0 \\ f_6 = 0 \end{cases} \qquad (4.5)$$

where x_i $(i = 1, 2, \ldots, 6)$ is the ith element of **X**.

Based on the definition of the state equation, state transition matrix and the procedure of the state estimation algorithm shown in Table 3.1, the parameters, especially the resistance, can be identified in real-time. Therefore, battery SOH can be calculated with the estimated resistance. For the adaptive state estimation methods, they can be implemented in two ways: joint estimation; and dual estimation.

The joint estimation methods identify battery model parameters first and then estimate the SOC in sequence. The dual estimation methods estimate battery model parameters and SOC on the same or different time scales. The joint estimation and dual estimation methods are indispensable because battery available capacity is gradually reduced due to its aging process in EVs. To obtain the accurate SOC, the currently available capacity of a battery needs to be estimated. Once the currently available capacity is determined, the SOH can be calculated by the ratio of the currently available capacity to the capacity of a fresh battery.

Figure 4.13 shows the basic frameworks for the joint estimation and dual estimation methods. Their differences can be explained as follows: first, the innovation sequence (the output prediction error) in the two methods is used differently, where the innovation sequence is the errors of battery terminal voltage. In the dual estimators, the voltage errors can be controlled and regulated in the two estimators. They coordinate each other to minimize the voltage error of each estimator. The dual estimators involve a closed-loop and feedback process. In the joint estimator, the voltage errors in the two estimators are separate and irrelevant. The joint estimators involve an open-loop process. Secondly, the relationship between the parameter estimation and state estimation is different. In the dual estimators, the state and parameter estimation processes will interact with each other. In the joint estimators, the state and parameter estimation processes work independently. The detailed computational procedure will be introduced in Sections 4.4 and 4.5.

We can also combine two or more methods to achieve a more reliable SOH estimation. The OCV and Ah counting methods can be combined through a battery model, and the pulse discharge/charge resistance, voltage characteristic and OCV methods can also be combined by using a battery model and an adaptive filtering method. The design requirement for the SOH indicator and the computational capability of the microprocessor will determine which of the combined methods should be selected. In the following, we introduce a combined method with solid electrolyte interphase (SEI) resistance and RLS filtering methods as an example.

In order to realize the online identification of the SEI resistance, a double resistance–capacitor (RC) network-based impedance model is selected, as shown in Figure 4.14.

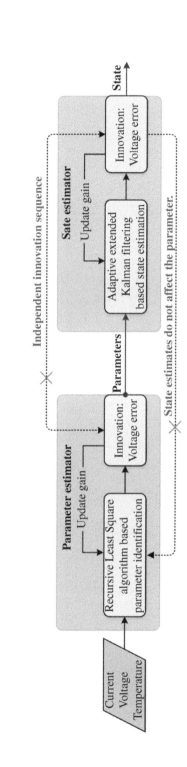

Figure 4.13 Frameworks for (a) joint estimation and (b) dual estimation of battery parameters and states.

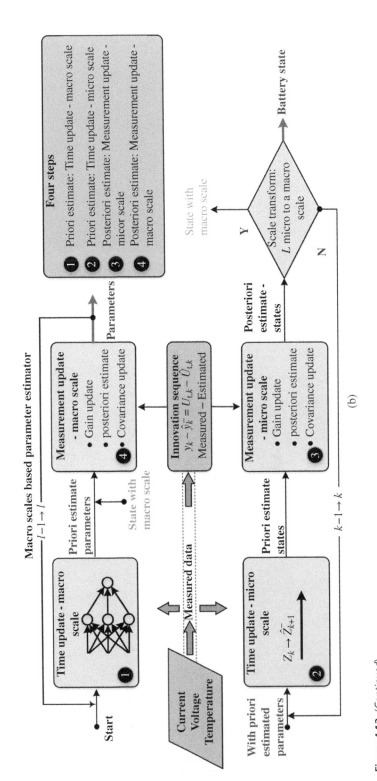

Figure 4.13 (*Continued*)

Figure 4.14 Double RC network-based impedance model.

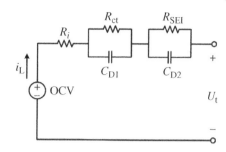

R_{ct} and C_{D1} in parallel account for the charge transfer progress, and they denote the charge transfer resistance and the double layer capacitance on the electrode, respectively. R_{SEI} and C_{D2} correspond to the SEI on the anode and they denote the SEI resistance and double layer capacitance on SEI, respectively [9]. It can be seen that the impedance model is very similar to the DP model shown in Section 2.3. Thus, we can use the online parameter identification method shown in Section 2.6 to identify the model parameters of the impedance model. The identified SEI resistance is shown in Figure 4.15.

The aging tests are carried out and the capacity profile is plotted in Figure 4.16. The capacity of the degraded cell decreases monotonically with increase in the cycle number. It is worth noting that the capacity decreases significantly between 400 and 550 cycles.

The correlation between SEI resistance and available capacity is plotted in Figure 4.17. It is shown that the average SEI resistance has a monotonic relationship with available capacity. The same trend is also observed in other batteries. To establish such a relationship, the fitting function is constructed in such a way that it will pass point (1, 1) as shown in Figure 4.18 and can be expressed as

$$R_{sei} = ae^{bQ_e} \tag{4.6}$$

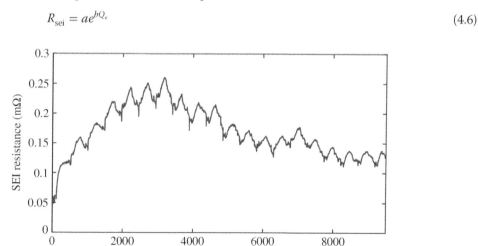

Figure 4.15 Identified SEI resistance.

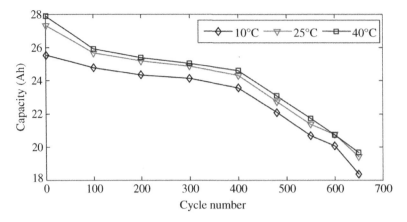

Figure 4.16 Capacity degradation of a LiB cell.

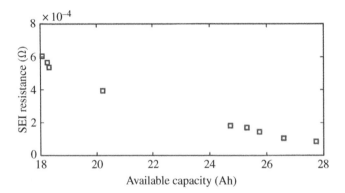

Figure 4.17 Correlation between SEI resistance and available capacity.

Table 4.1 Results of fitting function.

	a	b	SSE	RMSE	R^2
Values	698	−6.567	0.2056	0.1714	0.9933

where Q_e represents the capacity retention rate, and R_{sei} denotes the growth rate of SEI resistance, and a and b are the fitting parameters. Table 4.1 shows the fitting parameters, the sum of squares due to error (SSE) and root mean squared error (RMSE) for the fitting function. The identified parameters of the fitting function are substituted into Eq. (4.6) to obtain the calculated results. The comparison between the results obtained by the empirical equation or fitting model and those obtained by experiment has shown that there is good agreement, as shown in Figure 4.18.

Table 4.2 compares the experimental SOH with the model SOH. The SOH estimation errors are all less than 3%. It shows that the SEI resistance method can be combined well with RLS filtering methods.

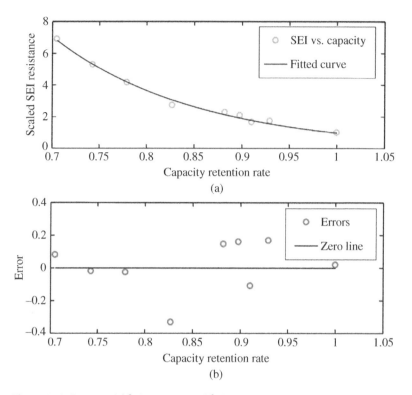

Figure 4.18 Exponential fitting curves and fitting errors.

Table 4.2 Comparison between experimental SOH and model SOH.

SOH	Estimated value	Relative error (%)
1	1.0	0
0.936	0.933	0.32
0.914	0.897	1.88
0.898	0.876	2.48
0.723	0.706	2.32
0.662	0.679	2.65

4.3.2 Data-Driven Methods

The data-driven methods for the SOH estimation do not need a mathematical model to describe the physical and chemical process inside a battery. Instead, they establish a model to describe degradation evolution of batteries only dependent on historical degradation data. Generally, they can be categorized as empirical and fitting methods, response surface-based optimization algorithms, and sample entropy methods.

4.3.2.1 Empirical and Fitting Methods

Empirical and fitting methods use available degradation data to estimate the lifetime of LiBs without the detailed knowledge of electrochemical cell design and material characteristic. Polynomial, exponential, power law, logarithmic, and trigonometric functions are commonly used as empirical and fitting models. The advanced intelligent algorithms and machine learning methods are also widely used for battery life prediction and SOH estimation. The simplicity of empirical or fitting models generally enables faster computations. The Arrhenius dynamics equation characterizes the temperature-dependent rate of a chemical reaction. It is often applied to model the temperature-induced alteration of diffusion coefficients, creep rates, and other thermal processes. The Arrhenius equation is also used to delineate the temperature-dependent degradation rates of electrochemical cells. The basic equation is as follows [10]:

$$\frac{dC}{dn} = \Lambda e^{-\frac{\Delta E}{R_g T}} = \Lambda e^{-\frac{\lambda}{T}} \tag{4.7}$$

where dC/dn is the changing rate of cell capacity with respect to aging cycle, Λ is the pre-exponential factor, and R_g is the universal gas constant (i.e. 8.314 J mol^{-1} K^{-1}). ΔE and T are the activation energy (in J mol^{-1}) and absolute temperature (in K), respectively. Λ and $\lambda = \Delta E/R_g$ are two unknown parameters that need to be calibrated. Integrating both sides of Eq. (4.7) over the cell lifespan yields:

$$C_r = -\Lambda n_c e^{-\frac{\lambda}{T}} \tag{4.8}$$

where C_r is the capacity reduction threshold indicative of the cell failure, and n_c is the cycle life. Two different temperatures, T_1 and T_2 ($T_1 > T_2$), are substituted into Eq. (4.8), giving

$$\Delta n_c = \frac{C_r}{\Lambda} \left(e^{-\frac{\lambda}{T_2}} - e^{-\frac{\lambda}{T_1}} \right) \tag{4.9}$$

where Δn_c is the lifetime deviation. It is evident that Eq. (4.9) is a quantitative description of the effect of the temperature gradient on cell life. The parameters of the Arrhenius equation are identified in Yang et al. [10] and then used to develop an Arrhenius equation-based cell-health assessment for EVs.

In addition to the Arrhenius equation, the cycle-induced capacity fade of a LFP battery with lots of experimental data is investigated in Wang et al. [11], and the cycle-life models are established based on the results from a large cycle-test matrix which includes three important parameters: temperature (-30 to $60\,°C$); DOD (90–10%); and discharge rate (C-rate, ranging from C/2 to 10C). It is demonstrated that the capacity fade follows a power law relationship with charge throughput between $15\,°C$ and $60\,°C$. Then, a simple battery life model that accounts for Ah-throughput (time), C-rates and temperature is established with good agreement with experimental data. For each discharge rate, a least squares regression model, which combines various factors (e.g. time, temperature, DOD) into an Ah-throughput dependent aging expression, is established for battery SOH prediction, as shown in Table 4.3, where Ah-throughput is defined as: number of cycles $\times$ DOD $\times$ full cell capacity [12].

4.3.2.2 Response Surface-Based Optimization Algorithms

There exists a monotonic relationship between battery OCV and SOC of a LiB. Such a relationship varies not only with operating temperatures but also with degradation

Table 4.3 Equations to predict capacity fade at a given discharge rate.

C-rate	Life model
C/2	$Q_{\text{loss}} = 30300 \cdot \exp\left(\dfrac{-31500}{RT}\right) \cdot (\text{Ah})^{0.552}$
2C	$Q_{\text{loss}} = 19330 \cdot \exp\left(\dfrac{-31000}{RT}\right) \cdot (\text{Ah})^{0.554}$
6C	$Q_{\text{loss}} = 12000 \cdot \exp\left(\dfrac{-29500}{RT}\right) \cdot (\text{Ah})^{0.56}$
10C	$Q_{\text{loss}} = 11500 \cdot \exp\left(\dfrac{-28000}{RT}\right) \cdot (\text{Ah})^{0.56}$

process of a battery indicated by the SOH. Therefore, it is important to establish the relationship between battery OCV, SOH (or available capacity) and SOC. Two steps are used to obtain their relationship [13].

First, we construct a three-dimensional response model of an OCV–SOC–capacity map covering the entire lifetime of a battery; it can be used to describe the battery capacity degradation characteristics and determine the corresponding SOC. The process of building a response surface model is as follows: (i) obtain the entire degradation data of battery capacity and OCV. (ii) Fit the OCV experimental data for discrete capacity points. It can be carried out with polynomials, the Nernst equation, and the Gaussian equation. (iii) Interpolate OCV–SOC curves two-dimensionally for different capacity points and establish the relationship between model parameters and capacity.

Secondly, a genetic algorithm (GA) is applied to identify battery capacity and initial SOC based on a battery model. Figure 4.19 shows a three-dimensional response surface-based OCV model of the battery under the entire degradation cycles. It is very time-consuming to obtain the OCV–SOC–capacity curve in the whole life cycle, but this curve is very important for BMSs in EVs which can be used to accurately predict battery capacity and SOC through the OCV.

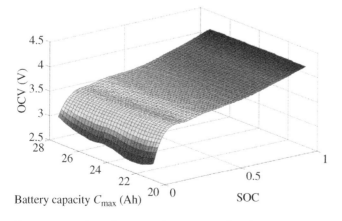

Figure 4.19 A three-dimensional response surface-based battery OCV model.

In the process of model parameter identification, the changes of the model parameters are reflected by the differences (or errors) of the terminal voltages between the estimation and experimental results. Different combinations of model parameters provide different terminal voltage errors and a set of parameters which can minimize the root mean value of the terminal voltage errors is the optimal set of model parameters which leads to the optimal battery model. Take the Thevenin model as an example, the terminal voltage is expressed as

$$U_t = U_{oc}(z, C_{max}) - U_D + i_L R_i \tag{4.10}$$

where C_{max} denotes the maximum available capacity of the battery. The OCV U_{oc} is a function with two independent variables, namely battery SOC (i.e. z) and C_{max}. Thus, the difference in the available capacity values reflected in the OCV–SOC curve can ultimately be reflected in the difference of terminal voltages. The available capacity of the battery can be incorporated into the model parameters. The initial value of SOC is also included in the parameters to be identified. At each sampling time, the SOC in EV driving conditions can be obtained by an Ah counting method. As a result, the estimation of the available capacity (or SOH) can be changed into the matching problem of searching for the best OCV–SOC relation on the OCV–SOC–capacity map. To solve this matching problem, we define battery parameters as

$$\hat{\theta}_{op} = \begin{bmatrix} z_0 & R_i & R_D & C_D & C_{max} \end{bmatrix} \tag{4.11}$$

The GA method is chosen to minimize the root mean square error of the terminal voltages which are considered as the fitness function in Eq. (4.12). The SOC at any time can be calculated by using the Ah counting method and the currently available capacity of the battery through the identified results.

$$\begin{cases} \min\{f(\hat{\theta}_{op})\} \\ f(\hat{\theta}_{op}) = \sqrt{\dfrac{\sum\limits_{k=1}^{N} [U_{t,k} - \hat{U}_{t,k}(\hat{\theta}_{op})]^2}{N}} \end{cases} \tag{4.12}$$

The lithium nickel manganese cobalt oxide (NMC) battery with nominal capacity of 25 Ah is used to show the response surface method. The actual available capacity of this battery at 25 °C is 27.31 Ah and its initial SOC is 82.7%. After the optimization method is applied, the results are obtained and are listed in Table 4.4. It can be seen that the estimated initial SOC is 84.3%. Comparing with the true initial SOC, the relative error

Table 4.4 Identified parameters by GA.

	z_{ini} (%)	R_i (mΩ)	R_D (mΩ)	C_D (F)	C_{max} (Ah)
Identified results	84.3	2.1	2.3	2.28e4	27.13

is only 1.6%. The estimated error of available capacity is 0.18 Ah and the relative error is 0.66%. Therefore, the response surface-based optimization algorithm can be used to accurately estimate the available capacity for the SOH calculation.

4.3.2.3 Sample Entropy Methods

The sample entropy (SampEn) in combination with a support vector machine were used to estimate battery SOH in Widodo et al. [14]. SampEn can provide computational means for assessing the predictability of a time series and it can also quantify the regularity of a data sequence. Therefore, when it is applied to analyze discharge voltage data, it could provide an indicator for battery health state. Given N data points from a time series $\{x(n)\} = x(1), x(2), \cdots, x(N)$, the SampEn algorithm is summarized as follows:

(a) Form vector sequences of size m, $X_m(1), \ldots, X_m(N - m + 1)$, defined by $X_m(i) = \{x(i), x(i + 1), \cdots x(i + m - 1)\}$, for $1 \leq i \leq N - m + 1$. These vectors represent m consecutive x values, starting with the ith point.

(b) Define the distance between vectors $X_m(i)$ and $X_m(j)$, $d|X_m(i), X_m(j)|$, as the absolute maximum difference between their scalar components:

$$d|X_m(i), X_m(j)| = \max_{k=0,\cdots,m-1}(|X(i+k) - X(j+k)|) \tag{4.13}$$

(c) For a given $X_m(i)$, count the number of $j(1 \leq j \leq N - m, j \neq i)$, denoted as B_i, such that the distance between $X_m(i)$ and $X_m(j)$ is less than or equal to r. Then, for $1 \leq i \leq N - m$:

$$B_i^m(r) = \frac{1}{N - m - 1} B_i \tag{4.14}$$

(d) Define $B^m(r)$ as:

$$B^m(r) = \frac{1}{N - m} \sum_{i=1}^{N-m} B_i^m(r) \tag{4.15}$$

(e) Increase the dimension to $m + 1$ and calculate A_i as the number of $X_{m+1}(i)$ within r or $X_{m+1}(j)$, where j ranges from 1 to $N - m$ $j \neq i$. Then, $A_i^m(r)$ is defined as:

$$A_i^m(r) = \frac{1}{N - m - 1} A_i \tag{4.16}$$

(f) Set $A^m(r)$ as:

$$A^m(r) = \frac{1}{N - m} \sum_{i=1}^{N-m} A_i^m(r) \tag{4.17}$$

(g) Finally, SampEn can be obtained as:

$$\text{SampEn}(m, r, N) = - \ln \left[\frac{A^m(r)}{B^m(r)} \right] \tag{4.18}$$

In Widodo et al. [14], it was introduced by utilizing machine learning methods, namely support vector machine and relevance vector machine. SampEn is employed as data input, while estimated SOH is employed as the target vector of learning algorithms.

4.4 Joint Estimation Method

4.4.1 Relationship Between SOC and Capacity

The SOC is defined as

$$z(t_2) = z(t_1) + \frac{1}{C_{max}} \int_{t_1}^{t_2} \frac{\eta i(\tau)}{3600} d\tau \qquad (4.19)$$

where $z(t_2)$ and $z(t_1)$ represent SOC at times t_2 and t_1, respectively, C_{max} represents the maximum available capacity of the battery, and η represents the coulomb efficiency ($\eta \approx 1$ as the self-discharge rate of the lithium battery is low). Equation (4.19) can be rewritten as

$$C_{max} = \int_{t_1}^{t_2} \frac{\eta i(\tau)}{3600} d\tau / [z(t_2) - z(t_1)] \qquad (4.20)$$

From Eq. (4.20), the maximum available capacity of the battery can be obtained using two SOCs at times t_2 and t_1 as well as the accumulated Ah during the period between t_2 and t_1. Figure 4.20 shows the procedure to calculate the maximum available capacity of the battery using Eq. (4.20).

Since the temperature has a large effect on battery capacity and the calculation of the battery available capacity in the short discharge interval causes a large fluctuation, a certain boundary condition is set for the capacity estimation. Generally, the temperature variation range is set as $\Delta T \leq \pm 5°C$, and the SOC variation range is set as $\Delta SOC \geq 20\%$. Furthermore, high-precision and high sampling frequency current sensors are used to improve the accuracy of charge accumulation. To obtain an accurate SOC estimation, the joint capacity and SOC estimation based on an H infinity state observer is introduced with the assumption of the known OCV–SOC relationship; the details can be found in Chapter 3. The flow chart for capacity estimation is shown in Figure 4.21.

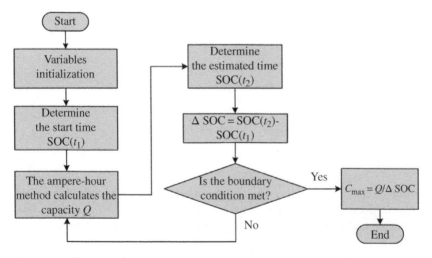

Figure 4.20 Flowchart for available capacity estimation based on SOC definition.

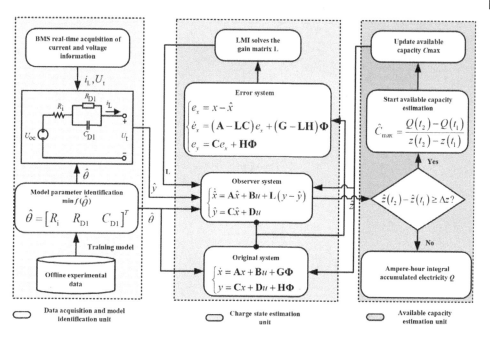

Figure 4.21 Flowchart of capacity estimation algorithm by joint estimator.

4.4.2 Case Study

The lithium manganese oxide (LMO) LiB cell introduced in Section 2.5.2 is used as an example to demonstrate the application of joint estimation. Based on the Thevenin model, the battery SOC is obtained by the H infinity observer, and then the obtained SOC is used to estimate the battery capacity. It is worth noting that the initial SOC is set as 20% away from the true SOC and the capacity estimation will start when ΔSOC is larger than 25%. Figure 4.22 indicates the estimation results of the SOC under the Dynamic Stress Test (DST) data. It can be seen that the estimated SOC can converge to

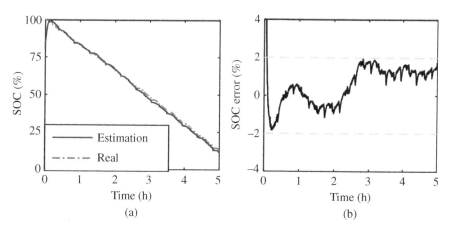

Figure 4.22 (a) Estimated battery SOC and (b) the error with the H infinity observer.

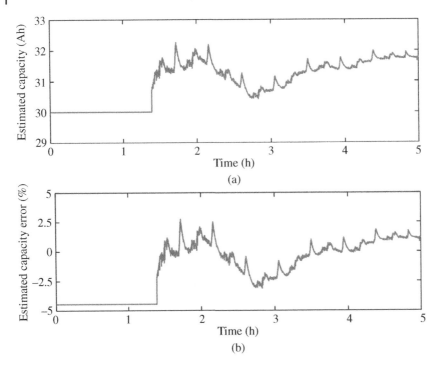

Figure 4.23 (a) Capacity estimation of battery and (b) estimation error.

the real value quickly with its error within ±2%, which shows that the estimated SOC accuracy is independent of the initial SOC and the initial capacity. Figure 4.23 shows the estimated capacity and the corresponding error. It can also be seen that the estimated capacity based on the H infinity observer is accurate with its error within 5%.

4.5   Dual Estimation Method

Generally, system parameters change slowly over time while system states are prone to change fast over time; a multi-time scale AEKF algorithm is used to estimate the system parameters in the macro time scale and the system state in the micro time scale [15]. For a system with fast and slow time varying characteristics, if the system states and parameters are estimated in the same time scale, the system parameters will change too frequently, reducing the stability of the estimation algorithms and increasing the computation burden. The multi-time scale approach can achieve more accurate and reliable estimation of the system parameters at low computational cost. The basic schematic is shown in Figure 4.13.

4.5.1   Implementation with the AEKF Algorithm

To reduce the complexity of a symbol system, the symbols in this section may differ from those of other chapters. Equation (4.21) is a nonlinear discrete system, which describes the multi-time scale system that contains the implicit state χ and the parameter θ.

$$\begin{cases} \chi_{k,l+1} = \mathbf{F}(\chi_{k,l}, \theta_k, \mathbf{u}_{k,l}) + \omega_{k,l}, \theta_{k+1} = \theta_k + \rho_k \\ \mathbf{Y}_{k,l} = \mathbf{G}(\chi_{k,l}, \theta_k, \mathbf{u}_{k,l}) + \upsilon_{k,l} \end{cases} \tag{4.21}$$

where $\chi_{k,l}$ is the states of the system when $t_{k,l} = t_{k,0} + l \times \Delta t$. k and l are double time scales. k describes the macro time scale while l describes the micro time scale. $\mathbf{u}_{k,l}$, which is the control matrix, is an input information of the system when the time is $t_{k,l}$. $\mathbf{Y}_{k,l}$ is the measurement matrix when the time is $t_{k,l}$. $\omega_{k,l}$ and ρ_k are the white noises of the system states and parameters, and their covariance matrices are $\mathbf{Q}_{k,l}^{\chi}$ and $\mathbf{Q}_k^{\theta}$. $\upsilon_{k,l}$ is the white noise of measurement, and its covariance matrix is $\mathbf{R}_{k,l}$. It is noted that not all parameters of a system need to be estimated in the macro time scale. The choice of time scales is based on the actual system. For the characteristics of a battery system, the states of a system are estimated in the micro time scale and the parameters of a system are estimated in the macro time scale. Figure 4.24 shows a schematic diagram of the multi-time scale AEKF, where AEKF$_\chi$ is the state observer in the micro time scale and AEKF$_\theta$ is the parameter observer in the macro time scale. Double AEKFs are used to carry out the multi-time scale estimation of states and parameters.

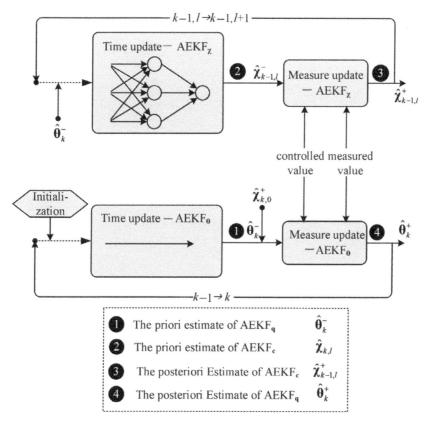

Figure 4.24 Schematic diagram of multi-time scale AEKF.

Based on the system described by Eq. (4.21), the calculation steps of the multi-time scale AEKF algorithm are summarized in the following.

Initialization of the algorithm: set the initial parameters of AEKF and $AEKF_\chi$ as

$$\boldsymbol{\theta}_0, \mathbf{P}_0^\theta, \mathbf{Q}_0^\theta, \mathbf{R}_0, \boldsymbol{\chi}_{0,0}, \mathbf{P}_{0,0}^\chi, \mathbf{Q}_{0,0}^\chi, \mathbf{R}_{0,0} \tag{4.22}$$

where $\boldsymbol{\theta}_0$, $\mathbf{P}_0^\theta$, and $\mathbf{Q}_0^\theta$ are the initial parameters, the initial parameter estimation error covariance matrix, and the initial system noise covariance matrix of AEKF. $\boldsymbol{\chi}_{0,0}$, $\mathbf{P}_{0,0}^\chi$, and $\mathbf{Q}_{0,0}^\chi$ are the initial system states, the initial parameter estimation error covariance matrix, and the initial system noise covariance matrix of AEKF. $\mathbf{R}_0$ and $\mathbf{R}_{0,0}$ are the measurement noise covariance and $\mathbf{R}_k = \mathbf{R}_{k,0:Lz-1}$, and L_z is the scale-switch transition value which is equal to the ratio of the macro time scale to the micro time scale, so $\mathbf{R}_0 = \mathbf{R}_{0,0}$. When the estimation starts, the values of time (0) change to values of time $(k-1)$, and the values of time (0, 0) change to values of time $(k-1, l-1)$.

For $k \in \{1, \cdots, \infty\}$, we calculate

$$\hat{\boldsymbol{\theta}}_k^- = \hat{\boldsymbol{\theta}}_{k-1}, \mathbf{P}_k^{\theta,-} = \mathbf{P}_{k-1}^\theta + \mathbf{Q}_{k-1}^\theta \tag{4.23}$$

Step 1. Time-update equations for parameters filter $\hat{\boldsymbol{\theta}}_k^-$ in the macro time scale (priori estimate).

For $l \in \{1, \cdots, Lz\}$, we calculate the state filter at each micro scale.

Step 2. Time-update equations for state $\hat{\boldsymbol{\chi}}_{k-1,l}^-$ in the micro time scale (priori estimate).

$$\hat{\boldsymbol{\chi}}_{k-1,l}^- = \mathbf{F}(\hat{\boldsymbol{\chi}}_{k-1,l-1}, \hat{\boldsymbol{\theta}}_k^-, \mathbf{u}_{k-1,l-1}), \quad \mathbf{P}_{k-1,l}^{\chi,-} = \mathbf{A}_{k-1,l-1}\mathbf{P}_{k-1,l-1}^\chi\mathbf{A}_{k-1,l-1}^\mathrm{T} + \mathbf{Q}_{k-1,l-1}^\chi \tag{4.24}$$

Step 3. Measurement-update equations for state $\hat{\boldsymbol{\chi}}_{k-1,l}^+$ in the micro time scale (posteriori estimate), for time-series calculation $l = 1 : L_z$ $(l \to L_z)$, we calculate the states estimation innovation matrix update by

$$\mathbf{e}_{k-1,l} = \mathbf{Y}_{k-1,l} - \mathbf{G}(\hat{\boldsymbol{\chi}}_{k-1,l}^-, \hat{\boldsymbol{\theta}}_k^-, \mathbf{u}_{k-1,l}) \tag{4.25}$$

The Kalman gain matrix update by

$$\mathbf{K}_{k-1,l}^\chi = \mathbf{P}_{k-1,l}^{\chi,-}(\mathbf{C}_{k-1,l}^\chi)^\mathrm{T}(\mathbf{C}_{k-1,l}^\chi\mathbf{P}_{k-1,l}^{\chi,-}(\mathbf{C}_{k-1,l}^\chi)^\mathrm{T} + \mathbf{R}_{k-1,l-1})^{-1} \tag{4.26}$$

The adaptive covariance matching (the voltage estimation error window function) update by

$$\mathbf{H}_{k-1,l}^\chi = \frac{1}{M_\chi}\sum_{i=l-M_\chi+1}^{l} \mathbf{e}_{k-1,i}\mathbf{e}_{k-1,i}^\mathrm{T} \tag{4.27}$$

The noise covariance update by

$$\mathbf{R}_{k-1,l} = \mathbf{H}_{k-1,l}^\chi - \mathbf{C}_{k-1,l}^\chi\mathbf{P}_{k-1,l}^{\chi,-}(\mathbf{C}_{k-1,l}^\chi)^\mathrm{T}, \mathbf{Q}_{k-1,l}^\chi = \mathbf{K}_{k-1,l}^\chi\mathbf{H}_{k-1,l}^\chi(\mathbf{K}_{k-1,l}^\chi)^\mathrm{T} \tag{4.28}$$

The system state estimation correction

$$\hat{\boldsymbol{\chi}}_{k-1,l}^+ = \hat{\boldsymbol{\chi}}_{k-1,l}^- + \mathbf{K}_{k-1,l}^\chi[\mathbf{Y}_{k-1,l} - \mathbf{G}(\hat{\boldsymbol{\chi}}_{k-1,l}^-, \hat{\boldsymbol{\theta}}_k^-, \mathbf{u}_{k-1,l})] \tag{4.29}$$

The state estimation error covariance update by

$$\mathbf{P}_{k-1,l}^{\chi,+} = (\mathbf{I} - \mathbf{K}_{k-1,l}^\chi\mathbf{C}_{k-1,l}^\chi)\mathbf{P}_{k-1,l}^{\chi,-} \tag{4.30}$$

Time scale transform when $l = L_z$.

$$\widehat{\chi}_{k,0}^{+} = \widehat{\chi}_{k-1,L_z}^{+}, \quad P_{k,0}^{\chi,+} = P_{k-1,L_z}^{\chi,+}, \quad Y_{k,0} = Y_{k-1,L_z}, \quad u_{k,0} = u_{k-1,L_z} \tag{4.31}$$

So, the calculations at the micro time scale for L_z times equivalent to one macro time scale have been completed.

Step 4. Measurement-update equations for state filter $\widehat{\theta}_k^{+}$ in the macro time scale (posteriori estimate).

The status estimation innovation matrix update by

$$e_k^{\theta} = Y_{k,0} - G(\widehat{\chi}_{k,0}^{+}, \widehat{\theta}_k^{-}, u_{k,0}) \tag{4.32}$$

The Kalman gain matrix update by

$$K_k^{\theta} = P_k^{\theta,-}(C_k^{\theta})^{T}(C_k^{\theta}P_k^{\theta,-}(C_k^{\theta})^{T} + R_{k-1})^{-1} \tag{4.33}$$

The adaptive covariance matching (the voltage estimation error window function) update by

$$H_k^{\theta} = \frac{1}{M_{\theta}} \sum_{i=k-M_{\theta}+1}^{k} e_i^{\theta}(e_i^{\theta})^{T} \tag{4.34}$$

The noise covariance update by

$$R_k = H_k^{\theta} - C_k^{\theta}P_k^{\theta,-}(C_k^{\theta})^{T}, Q_k^{\theta} = K_k^{\theta}H_k^{\theta}(K_k^{\theta})^{T} \tag{4.35}$$

The system state estimation correction

$$\widehat{\theta}_k^{+} = \widehat{\theta}_k^{-} + K_k^{\theta}e_k^{\theta} \tag{4.36}$$

The state estimation error covariance update

$$P_k^{\theta,+} = (I - K_k^{\theta}C_k^{\theta})P_k^{\theta,-} \tag{4.37}$$

where

$$A_{k-1,l-1} = \left.\frac{\partial F(\chi, \widehat{\theta}_k^{-}, u_{k-1,l-1})}{\partial \chi}\right|_{\chi=\widehat{\chi}_{k-1,l-1}}, C_{k-1,l}^{\chi} = \left.\frac{\partial G(\chi, \widehat{\theta}_k^{-}, u_{k-1,l})}{\partial \chi}\right|_{\chi=\widehat{\chi}_{k-1,l}},$$

$$C_k^{\theta} = \left.\frac{\partial G(\widehat{\chi}_{k,0}, \theta, u_{k,0})}{\partial \theta}\right|_{\theta=\widehat{\theta}_k^{-}} \tag{4.38}$$

where C_k^{θ} is computed as

$$C_k^{\theta} = \left.\frac{\partial G(\widehat{\chi}_{k,0}, \theta, u_{k,0})}{\partial \theta}\right|_{\theta=\widehat{\theta}_k^{-}} = \frac{\partial G(\widehat{\chi}_{k,0}, \widehat{\theta}_k^{-}, u_{k,0})}{\partial \widehat{\theta}_k^{-}} + \frac{\partial G(\widehat{\chi}_{k,0}, \widehat{\theta}_k^{-}, u_{k,0})}{\partial \widehat{\chi}_{k,0}}\frac{d\widehat{\chi}_{k,0}}{d\widehat{\theta}_k^{-}} \tag{4.39}$$

Considering $\widehat{\chi}_{k,0} = \widehat{\chi}_{k-1,L_z}^{+}$, according to Eq. (4.29), it gives

$$\frac{d\widehat{\chi}_{k,0}}{d\widehat{\theta}_k^{-}} = \frac{d\widehat{\chi}_{k-1,L_z}^{+}}{d\widehat{\theta}_k^{-}}$$

$$= \frac{d}{d\widehat{\theta}_k^{-}}(\widehat{\chi}_{k-1,L_z-1}^{-} + K_{k-1,L_z-1}^{\chi}(Y_{k-1,L_z-1} - G(\widehat{\chi}_{k-1,L_z-1}^{-}, \widehat{\theta}_k^{-}, u_{k-1,L_z-1}))) \tag{4.40}$$

$$\frac{d}{d\widehat{\theta}_k^-}(\mathbf{K}_{k-1,L_z-1}^{\chi}Y_{k-1,L_z-1}) = Y_{k-1,L_z-1}\frac{\partial\mathbf{K}_{k-1,L_z-1}^{\chi}}{\partial\widehat{\theta}_k^-} \tag{4.41}$$

$$\frac{d}{d\widehat{\theta}_k^-}(\mathbf{K}_{k-1,L_z-1}^{\chi}\mathbf{G}(\widehat{\chi}_{k-1,L_z-1}^-, \widehat{\theta}_k^-, \mathbf{u}_{k-1,L_z-1}))$$

$$= \mathbf{K}_{k-1,L_z-1}^{\chi}\frac{\partial\mathbf{G}(\widehat{\chi}_{k-1,L_z-1}^-, \widehat{\theta}_k^-, \mathbf{u}_{k-1,L_z-1})}{\partial\widehat{\theta}_k^-} + \frac{\partial\mathbf{K}_{k-1,L_z-1}^{\chi}}{\partial\widehat{\theta}_k^-}\mathbf{G}(\widehat{\chi}_{k-1,L_z-1}^-, \widehat{\theta}_k^-, \mathbf{u}_{k-1,L_z-1}) \tag{4.42}$$

Herein, the multi-time scale AEKF estimation of states and parameters is completed. Then, the time of the state estimation filter has been changed from $t(k)^+$ to $t(k+1)^-$ which will prepare the estimation at the $(k+1)$th time and make $\widehat{\chi}_{k,0} = \widehat{\chi}_{k,0}^+$ and $\widehat{\theta}_k = \widehat{\theta}_k^+$.

4.5.2 SOC–SOH Estimation

According to the state equation of the Thevenin model, the battery system is transformed into a nonlinear discrete-time system as

$$\begin{bmatrix} U_{k,l}^{D} \\ z_{k,l} \end{bmatrix} = \begin{bmatrix} \exp\left(-\dfrac{\Delta t}{R_D C_D}\right) & 0 \\ 0 & 1 \end{bmatrix}\begin{bmatrix} U_{k,l-1}^{D} \\ z_{k,l-1} \end{bmatrix} + \begin{bmatrix} \left(1 - \exp\left(-\dfrac{\Delta t}{R_D C_D}\right)\right)R_D \\ -\dfrac{\eta_i \Delta t}{C_a} \end{bmatrix}i_{k,l} \tag{4.43}$$

where $U_{k,l}^{D}$ and $z_{k,l}$ are the polarization voltage and the SOC of a battery at time $t_{k,l}$, respectively, and $i_{k,l}$ is the battery current. Its system equation can be described as:

$$\begin{cases} \chi_{k,l+1} = \mathbf{F}(\chi_{k,l}, \theta_k, \mathbf{u}_{k,l}) = \begin{bmatrix} \exp\left(-\dfrac{\Delta t}{R_D C_D}\right) & 0 \\ 0 & 1 \end{bmatrix}\chi_{k,l} \\[20pt] \qquad + \begin{bmatrix} \left(1 - \exp\left(-\dfrac{\Delta t}{R_D C_D}\right)\right)R_D \\ -\dfrac{\eta_i \Delta t}{C_a} \end{bmatrix}\mathbf{u}_{k,l+1} \\[20pt] Y_{k,l} = \mathbf{G}(\chi_{k,l}, \theta_k, \mathbf{u}_{k,l}) = g(z_{k,l}, C_a) - U_{k,l}^{D} - R_o\mathbf{u}_{k,l} \end{cases} \tag{4.44}$$

where the state matrix, parameters matrix, control matrix and output matrix of the system are as follows:

$$\begin{cases} \chi_{k,l} = \begin{bmatrix} U_{k,l}^{D} & z_{k,l} \end{bmatrix}^{\mathrm{T}} \\ \theta_k = \begin{bmatrix} R_o & R_D & C_D & C_a \end{bmatrix}_k^{\mathrm{T}} \\ \mathbf{u}_{k,l} = i_{k,l} \\ Y_{k,l} = U_{k,l}^{\mathrm{t}} \end{cases} \tag{4.45}$$

where $U_{k,l}^t$ is the battery terminal voltage at time $t_{k,l}$. The state matrix and the observation matrix of the system are as follows:

$$\mathbf{A}_{k-1,l-1} = \begin{bmatrix} \exp\left(-\dfrac{\Delta t}{R_D C_D}\right) & 0 \\ 0 & 1 \end{bmatrix} \tag{4.46}$$

$$\mathbf{C}_{k-1,l}^{\chi} = \begin{bmatrix} -1 & \dfrac{\partial g_{k-1,l}(z, C_a)}{\partial z_{k-1,l}^{-}} \end{bmatrix} \tag{4.47}$$

$$\mathbf{C}_k^{\theta} = \begin{bmatrix} -i_{k,0} & 0 & 0 & \dfrac{\partial g(z, C_a)}{\partial z_{k,0}} \dfrac{\partial z_{k,0}}{\partial \widehat{C}_{a,k}^{-}} \end{bmatrix} + \mathbf{C}_{k,0}^{\chi}\left(\dfrac{\partial F(\widehat{\chi}_{k-1,L_z-1}^{-}, \widehat{\theta}_k^{-}, \mathbf{u}_{k-1,L_z-1})}{\partial \widehat{\theta}_k^{-}} \right.$$

$$\left. + \dfrac{\partial F(\widehat{\chi}_{k-1,L_z-1}^{-}, \widehat{\theta}_k^{-}, \mathbf{u}_{k-1,L_z-1})}{\partial \widehat{\chi}_{k-1,L_z-1}^{-}} \dfrac{d\widehat{\chi}_{k-1,L_z-1}^{-}}{d\widehat{\theta}_k^{-}} \right) \tag{4.48}$$

where

$$\dfrac{\partial g_{k-1,l}(z, C_a)}{\partial z_{k,l}} = \alpha_1 + 2\alpha_2 z_{k,l} + 3\alpha_3 (z_{k,l})^2 - \alpha_4/(z_{k,l})^2 + \alpha_5/z_{k,l} - \alpha_6/(1 - z_{k,l}) \tag{4.49}$$

where

$$\dfrac{\partial z_{k,0}}{\partial \widehat{Q}_{a,k}^{-}} = \dfrac{\eta_i i_{k,l-1} \Delta t}{(\widehat{C}_{a,k}^{-})^2} \tag{4.50}$$

$$\dfrac{\partial F(\widehat{\chi}_{k-1,L_z-1}^{-}, \widehat{\theta}_k^{-}, \mathbf{u}_{k-1,L_z-1})}{\partial \widehat{\theta}_k^{-}} = \begin{bmatrix} 0 & \sigma_1 & \sigma_2 & 0 \\ 0 & 0 & 0 & \sigma_3 \end{bmatrix} \tag{4.51}$$

where $\sigma_1, \sigma_2, \sigma_3$ are

$$\begin{cases} \sigma_{1,k} = \dfrac{\widehat{U}_{k-1,L_z-1}^{D} \Delta t}{R_D^2 C_D} \exp\left(-\dfrac{\Delta t}{R_D C_D}\right) - \dfrac{i_{k-1,L_z-1}\Delta t}{R_D C_D} \exp\left(-\dfrac{\Delta t}{R_D C_D}\right) \\ \qquad\quad -i_{k-1,L_z-1}\left(\exp\left(-\dfrac{\Delta t}{R_D C_D}\right) - 1\right) \\ \sigma_{2,k} = \dfrac{\widehat{U}_{k-1,L_z-1}^{D}\Delta t}{R_D C_D^2} \exp\left(-\dfrac{\Delta t}{R_D C_D}\right) - \dfrac{i_{k-1,L_z-1}\Delta t}{C_D^2} \exp\left(-\dfrac{\Delta t}{R_D C_D}\right) \\ \sigma_{3,k} = \eta_i i_{k-1,L_z-1}\Delta t/(\widehat{C}_{a,k}^{-})^2 \end{cases} \tag{4.52}$$

Figure 4.25 is the flowchart of a multi-time scale AEKF for the estimation of battery states and parameters.

In practical applications, when the current or power excitation is loaded into a battery, the data acquisition system will measure the current and voltage of the battery in

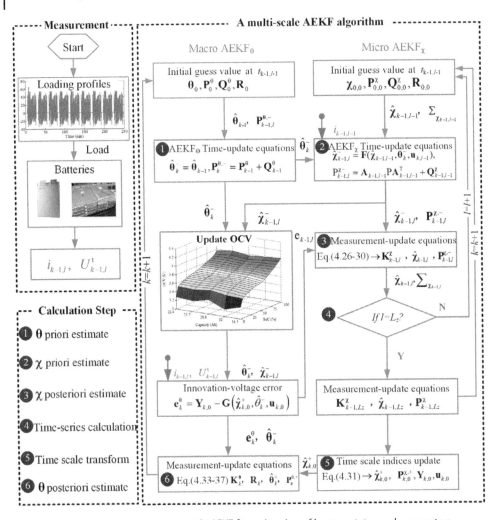

Figure 4.25 Flowchart of multi-time scale AEKF for estimation of battery states and parameters.

real-time and carry out the estimation of battery states and parameters. The detailed operation process is as follows:

Step 1. At each macro time scale, the macro AEKF_θ executes the time-update of parameters and calculates the priori parameter estimate $\hat{\boldsymbol{\theta}}_k^-$ and error covariance $\mathbf{P}_k^{\theta,-}$ by Eq. (4.23).

Step 2. After the time-update of AEKF_θ, the micro AEKF_χ executes the time-update of states with $\hat{\boldsymbol{\theta}}_k^-$ at each micro time scale, where the priori state estimate $\hat{\boldsymbol{\chi}}_{k-1,l}^-$ and its error covariance $\mathbf{P}_{k-1,l}^{\chi,-}$ are computed by Eq. (4.24). After the time-update of AEKF_θ and AEKF_χ, the OCV value is ready to update with the prior SOC and capacity estimates and then the predicted terminal voltage is also ready to update.

Step 3. At each micro time scale, the micro AEKF_χ executes the measurement-update followed by the time-update of AEKF_χ in step 2, where the posteriori state estimate $\widehat{\chi}^+_{k-1,l}$ and its error covariance $\mathbf{P}^{\chi,+}_{k-1,l}$ are calculated by Eqs. (4.25)–(4.30).

Step 4. After the posteriori state estimate, we compare the micro-scale l with the scale-switch transition value L_z. If the micro-scale l does not reach L_z, the state estimate $\widehat{\chi}^+_{k-1,l}$ will transmit to step 2 to serve as an initial value at time $t_{k-1,l}$ and then to estimate the state again. If l reaches L_z, the posteriori state estimate $\widehat{\chi}^+_{k-1,l}$ and its error covariance $\mathbf{P}^{\chi,+}_{k-1,l}$ are ready to update for the next macro time scale.

Step 5. We update all the micro time scale indices by Eq. (4.31), such as $\widehat{\chi}_{k-1,L} = \widehat{\chi}_{k,0}$. In other words, the estimated values at the time $t_{k-1,L}$ are ready to update to the values at $t_{k,0}$ for parameter estimation and the next state estimation.

Step 6. After the state estimates, the macro AEKF_θ is ready to execute the measurement-update, where posteriori parameter estimate $\widehat{\theta}^+_k$ and covariance $\mathbf{P}^{\theta,+}_k$ are computed by Eqs. (4.32)–(4.37).

After the above six steps, the estimation of battery parameters and states is achieved, the obtained parameters and states will serve as the initial values of the next estimation.

4.5.3 Case Study

The DST testing data of the NMC LiB cells is used as an example to illustrate the dual estimation of the SOC and capacity based on a multi-time scale AEKF algorithm. First, we show the estimation performance of a single-time scale AEKF algorithm, where the scale- switch transition value L_z is set as 1 second. Then, we show the estimation performance of a multi-time scale AEKF algorithm, where L_z is set as 60 seconds. The same conditions are applied to both cases with the initial SOC and initial available capacity set to 60% and 30 Ah, respectively.

Figure 4.26 shows the single-time scale AEKF estimation results under the reference SOC of 100% and true available capacity of 31.6 Ah. Figure 4.26a,c shows the terminal voltage estimation results and their estimation errors. Figure 4.26b,d shows the SOC estimation results and their estimation errors. Figure 4.26e,f shows the available capacity estimation results and their estimation errors.

It can be seen that the maximum voltage estimation error is less than 35 mV and the maximum SOC estimation error is less than 0.5% after convergence. The results show that a single-time scale AEKF algorithm can ensure the accurate estimation of the battery SOC and the available capacity. This indicates that the single-time scale method can obtain the accurate estimation of the parameters and states under inexact initial SOC value and available capacity. However, it can be observed from Figure 4.26 that the parameters or states estimation based on a single-time scale method are susceptible to uncertain conditions. For example, when the battery current fluctuates greatly, the terminal voltage and the estimated value of the available capacity are more volatile. Figure 4.26e shows the estimated capacity as a continuous fluctuation and every peak point appears at the time that the current value changes from large value to static state in the DST condition. Besides, there is a large amount of matrix calculation in this single-time scale AEKF algorithm and its calculation time takes about 4.709 seconds.

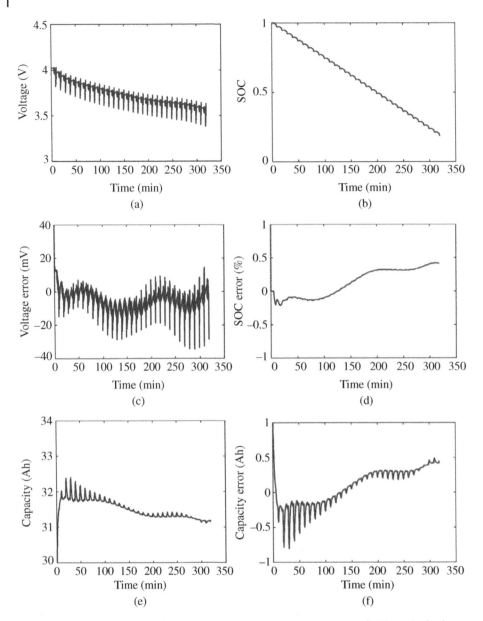

Figure 4.26 Capacity estimation with a single-time scale AEKF ($L_z = 1$ second): (a) terminal voltage; (b) reference SOC; (c) voltage error; (d) SOC error; (e) capacity estimation; (f) capacity estimation error.

Figure 4.27 shows the multi-time scale AEKF estimation results. Figure 4.27a,c shows the terminal voltage estimation results and their estimation errors. Figure 4.27b,d shows the SOC estimation results and their estimation errors. Figure 4.27e,f shows the available capacity estimation results and their estimation errors.

It can be seen that when a multi-time scale AEKF is used, the maximum voltage estimation error is less than 25 mV, the maximum SOC estimation error is less than

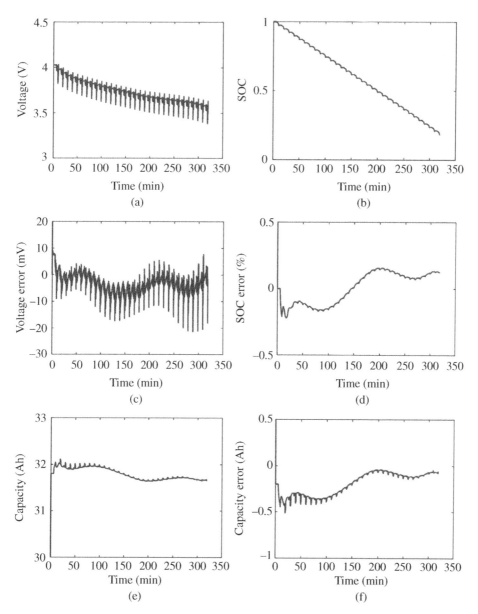

Figure 4.27 Capacity estimation with a multi-time scale AEKF ($L_z = 60$ seconds): (a) terminal voltage; (b) reference SOC; (c) voltage error; (d) SOC error; (e) capacity estimation; (f) capacity estimation error.

0.25%, and the maximum available capacity estimation error is less than 0.5 Ah after convergence. More importantly, the available capacity estimation result is more stable even over the time that the current value changes from large value to static state. The estimation results are converged to the reference value more quickly. The average calculation time is only about 2.512 seconds.

Table 4.5 SOC and capacity estimation and their estimation errors.

Capacity (Ah)	SOC estimate error (%)			Available capacity estimate error (Ah)			
	Max error	Average error	Standard deviation	Max error	Average error	Standard deviation	Error after convergence
31.6	0.424	0.215	0.120	0.678	0.055	0.086	<0.2
29.5	0.392	0.199	0.108	0.597	0.049	0.069	<0.2
25.7	0.673	0.756	0.204	0.843	0.061	0.071	<0.3

The estimation results based on different time scales show that: (i) when the multi-time scale AEKF is applied to estimate battery parameters and states, it provides a robust estimation algorithm which has a good ability of correction for the imprecise battery capacity and initial SOC value. (ii) The multi-time scale AEKF method can obtain higher estimation accuracy than the single-time scale one due to its capability of quick convergence to the reference value. After converging, the estimation errors of voltage, SOC and available capacity all are less than 1%. More importantly, the calculation time reduces from 4.709 to 2.512 seconds (by 47%), which is more suitable for the application of real-time BMSs for EVs.

It should be noted that the longer scale-switch transition value L_z does not mean higher accuracy for the estimation of the parameters and states of the battery. Analysis of the estimation results in different time scales is conducted. When L_z is too short, battery capacity estimation results can be easily affected by EV dynamic working conditions, leading to its large fluctuation. In reality, battery capacity changes slowly and does not affect the voltage estimation error over a short time, it is unnecessary to estimate battery capacity as frequent battery terminal voltage. When L_z is too long, battery SOC has largely changed which affects the parameters of the battery model; this can cause slow convergence of battery available capacity. As a result, it is found that the estimation accuracy is best when L_z is within the range of 60–90 seconds for most LiBs.

The estimation accuracy of a multi-time scale AEKF algorithm with different battery aging levels is also discussed for the same L_z value ($L_z = 60$ seconds). Battery SOC and available capacity estimation results after 10 minutes of calculation are shown in Table 4.5, where the initial available capacity is set at 30 Ah and the initial SOC value at 100%.

It can be seen that the maximum SOC estimation error is less than 1%, the maximum available capacity estimation error is less than 1 Ah, and the capacity estimation error is less than 0.3 Ah after convergence. Therefore, a multi-time scale AEKF algorithm can realize more accurate and robust estimation of battery available capacity and SOC under different aging levels, incorrect settings of initial battery available capacity and SOC.

4.6 Summary

This chapter systematically summarizes the state of the art of battery SOH estimation methods. Two groups of methods, experimental methods and model-based estimation methods, are discussed.

In the experimental methods, direct measurement methods and indirect analysis methods are explained. They store the whole degradation data of the battery and then analyze the influence of the main parameters on battery degradation, such as battery available capacity or resistances, which can be used to calculate the SOH. These methods are generally implemented in a laboratory environment.

In the model-based estimation methods, adaptive state estimation methods and data-driven methods are explored. They use advanced filtering techniques, state estimation methods and machine learning methods to analyze battery operational data and identify battery parameters that affect battery lifetime for the SOH calculation, which can reduce the dependency of performances on the operational data and achieve a more reliable estimation. In the adaptive state estimation methods, the dual and joint estimation methods are explored in detail. Joint estimation and dual estimation with a multi-time scale AEKF are suitable for a real-time BMS in EVs.

References

1 Berecibar, M., Gandiaga, I., Villarreal, I. et al. (2016). Critical review of state of health estimation methods of Li-ion batteries for real applications. *Renewable and Sustainable Energy Reviews* 56: 572–587.

2 Xiong, R., Tian, J., Mu, H., and Wang, C. (2017). A systematic model-based degradation behavior recognition and health monitoring method for lithium-ion batteries. *Applied Energy* https://doi.org/10.1016/j.apenergy.2017.05.124.

3 Waag, W., Fleischer, C., and Sauer, D.U. (2014). Critical review of the methods for monitoring of lithium-ion batteries in electric and hybrid vehicles. *Journal of Power Sources* 258: 321–339.

4 Eddahech, A., Briat, O., and Vinassa, J. (2014). Determination of lithium-ion battery state-of-health based on constant-voltage charge phase. *Journal of Power Sources* 258: 218–227.

5 Ma, Z., Wang, Z., Xiong, R., and Jiang, J. (2018). A mechanism identification model based state-of-health diagnosis of lithium-ion batteries for energy storage applications. *Journal of Cleaner Production* 193: 379–390.

6 Ma, Z., Jiang, J., Shi, W. et al. (2015). Investigation of path dependence in commercial lithium-ion cells for pure electric bus applications: aging mechanism identification. *Journal of Power Sources* 274: 29–40.

7 He, H., Xiong, R., and Guo, H. (2012). Online estimation of model parameters and state-of-charge of LiFePO$_4$ batteries in electric vehicles. *Applied Energy* 89 (1): 413–420.

8 Xiong, R., Sun, F., Gong, X., and Gao, C. (2014). A data-driven based adaptive state of charge estimator of lithium-ion polymer battery used in electric vehicles. *Applied Energy* 113: 1421–1433.

9 Christian, F., Wladislaw, W., Hans-Martin, H., and Dirk, U.S. (2014). On-line adaptive battery impedance parameter and state estimation considering physical principles in reduced order equivalent circuit battery models: Part 1. Requirements, critical review of methods and modeling. *Journal of Power Sources* 260: 276–291.

10 Yang, Y., Hu, X., Qing, D., and Chen, F. (2013). Arrhenius equation-based cell-health assessment: application to thermal energy management design of a HEV NiMH battery pack. *Energies* 6 (5): 2709–2725.

11 Wang, J., Liu, P., Hicks-Garner, J. et al. (2011). Cycle-life model for graphite-LiFePO$_4$ cells. *Journal of Power Sources* 196 (8): 3942–3948.

12 Jin, X., Vora, A.P., Hoshing, V., Saha, T., Shaver, G. M., Wasynczuk, O. and Varigonda S. (2017) Comparison of Li-ion battery degradation models for system design and control algorithm development. 2017 American Control Conference (ACC), Seattle, WA, 74–79.

13 Yang, R., Xiong, R., He, H. et al. (2017). A novel method on estimating the degradation and state of charge of lithium-ion batteries used for electrical vehicles. *Applied Energy* 207: 336–345.

14 Widodo, A., Shim, M.C., Caesarendra, W., and Yang, B.S. (2011). Intelligent prognostics for battery health monitoring based on sample entropy. *Expert Systems with Applications* 38 (9): 11763–11769.

15 Xiong, R., Sun, F., Chen, Z., and He, H. (2014). A data-driven multi-scale extended Kalman filtering based parameter and state estimation approach of lithium-ion polymer battery in electric vehicles. *Applied Energy* 113 (1): 463–476.

5

Battery State of Power Estimation

5.1 Background

The state of power (SOP) estimation of battery systems is indispensable to ensure the safe and reliable operation of electric vehicles (EVs). The SOP evaluates the maximum charge and discharge capability of battery systems. It can be used to estimate power requirements on accelerating and climbing, optimally balance power between the battery system and other power sources such as ultracapacitors, avoid overcharge or over-discharge and maximize the ability of regenerative braking.

In this chapter, instantaneous SOP estimation methods are discussed in Section 5.2, including the hybrid pulse power characterization (HPPC) method, the state of charge (SOC)-limited method, the voltage-limited method, and the multi-constrained dynamic (MCD) method. Then, the SOC-limited method, the voltage-limited method, and the MCD method are extended to continuous SOP estimation in Section 5.3; the adaptive extended Kalman filter (AEKF) and recursive least squares (RLS) are utilized to jointly estimate SOC, SOP, and model parameters in the presence of uncertainties of battery states and model parameters. Section 5.4 provides the summary.

5.2 Instantaneous SOP Estimation Methods

There are four instantaneous SOP (or peak power) estimation methods.

1. *The HPPC method* [1]. This method calculates instantaneously available current and power by taking the upper and lower voltage limits into consideration. It is a common method for battery management systems (BMSs) in EVs due to its simplicity.
2. *The SOC-limited method* [2]. This method provides an optimistic estimation of the peak current based on the allowable maximum and minimum SOCs. It is usually combined with other methods in real EV applications.
3. *The voltage-limited method* [3]. Similar to the HPPC method, the R_{int} model is used to calculate the SOP. In addition, this method also considers the change of open circuit voltage (OCV) during the sampling interval Δt, resulting in a more moderate SOP estimation.
4. *The MCD method* [2]. In this method, the real-time peak power of a battery system is calculated based on the multiple constraints, such as terminal voltage, current,

Advanced Battery Management Technologies for Electric Vehicles, First Edition. Rui Xiong and Weixiang Shen.
© 2019 John Wiley & Sons Ltd. Published 2019 by John Wiley & Sons Ltd.

SOC, and power. The influences of electrochemical kinetics, thermodynamics and relaxation effect on the peak power are also considered.

In the following, we introduce these four instantaneous SOP estimation methods.

5.2.1 HPPC Method

The HPPC method was developed by the Idaho National Engineering and Environmental Laboratory under the Partnership for New Generation Vehicles (PNGV) and is the most frequently used method [4]. It estimates the peak power of a battery cell based on the R_{int} model comprising an OCV and an internal resistance. The terminal voltage of each cell in a battery system is expressed as

$$U_{t,m}(t) = U_{oc,m}[z_m(t)] - R_i i_{L,m}(t) \tag{5.1}$$

where m refers to the mth cell, and t and z are time and SOC, respectively; $U_{oc,m}[z_m(t)]$ is the OCV at SOC $z_m(t)$ of the cell m; $i_{L,m}(t)$ is the load current of the cell m, and R_i is the internal resistance of each cell for charging (R_{chg}) or discharging (R_{dis}). Different values of R_i are taken for the charging or discharging process.

Considering the terminal voltage within the upper and lower limits ($U_{t,\,min} \leq U_{t,\,m}(t) \leq U_{t,\,max}$), the peak charge and discharge currents under the voltage constraints of the cell m are described as

$$\begin{cases} i_{min,m}^{chg,HPPC} = \dfrac{U_{oc,m} - U_{t,max}}{R_{chg}} \\[2mm] i_{max,m}^{dis,HPPC} = \dfrac{U_{oc,m} - U_{t,min}}{R_{dis}} \end{cases} \tag{5.2}$$

where $U_{t,max}$ and $U_{t,min}$ are the maximum terminal voltages during charging and the minimum terminal voltage during discharging, respectively. $i_{min,m}^{chg,HPPC}$ and $i_{max,m}^{dis,HPPC}$ are the minimum charge current and maximum discharge currents of the cell m. Then, the peak power of a battery system is calculated as

$$\begin{cases} P_{min}^{chg} = n_s n_p \max_m (U_{t,max} \times i_{min,m}^{chg,HPPC}) \\[2mm] P_{max}^{dis} = n_s n_p \min_m (U_{t,min} \times i_{max,m}^{dis,HPPC}) \end{cases} \tag{5.3}$$

where the battery system consists of n_s battery packs connected in series and each pack consists of n_p individual cells in parallel ($n_s \geq 1, n_p \geq 1$).

The HPPC method can be used to estimate absolute available peak power of a battery system. However, it cannot be used to estimate continuous available peak power at the next time interval Δt. Also, the R_{int} model cannot simulate battery dynamic characteristics, resulting in over-optimistic estimation of peak power which may lead to safety and health issues caused by overcharge and over-discharge.

This method has two drawbacks. First, it is not suitable for continuous peak current estimation that is available for the next sample interval Δt, and secondly, it ignores the limits of current, power, and SOC, which has an adverse effect on safety and the health of battery systems.

5.2.2 SOC-Limited Method

In the SOC-limited method, the maximum and minimum SOCs are used to calculate the peak currents of a battery system, which are then utilized to estimate available peak power. When a battery is discharged (or charged) at a constant current during a specified time period between t and $t + \Delta t$, the SOC of the cell m at the time $(t + \Delta t)$ is calculated by using ampere-hour (Ah) counting

$$z_m(t + \Delta t) = z_m(t) - i_{L,m}(t)\left(\frac{\eta_i \Delta t}{C_{max}}\right)$$

(5.4)

where η_i is a coulomb efficiency, which is a function of the current $i_{L,m}(t)$. C_{max} is the maximum available capacity.

In consideration of the upper and lower SOC limits z_{max} and z_{min}, respectively, the peak charge and discharge current of the cell m are then expressed as

$$\begin{cases} i_{min,m}^{chg,SOC} = \dfrac{z_m(t) - z_{max}}{\eta_i \Delta t / C_{max}} \\ i_{max,m}^{dis,SOC} = \dfrac{z_m(t) - z_{min}}{\eta_i \Delta t / C_{max}} \end{cases}$$

(5.5)

Then, the peak currents of the battery system are

$$\begin{cases} i_{min}^{chg,SOC} = \max_{m}(i_{min,m}^{chg,SOC}) \\ i_{max}^{dis,SOC} = \max_{m}(i_{max,m}^{dis,SOC}) \end{cases}$$

(5.6)

It is assumed that the SOC is accurate for every cell in the system. If this assumption is not satisfied, the peak charge and discharge current are approximately given by

$$\begin{cases} i_{min}^{chg,SOC} \approx \dfrac{z(t) - z_{max}}{\eta_i \Delta t / C_{max}} \\ i_{max}^{dis,SOC} \approx \dfrac{z(t) - z_{min}}{\eta_i \Delta t / C_{max}} \end{cases}$$

(5.7)

The SOC-limited method aims to estimate peak power in the time period Δt within the SOC limits. It gives over-optimistic estimations of the peak current if a battery is allowed to operate in a large SOC range.

5.2.3 Voltage-Limited Method

The HPPC method estimates an instantaneously available peak current, which is available from t to $t + \Delta t$. If Δt becomes larger, the battery is at risk of being over charged or over discharged. To solve this problem, the mathematical expression of the R_{int} model in Eq. (5.1) should be revised as

$$U_{t,m}(t + \Delta t) = U_{oc,m}[z_m(t + \Delta t)] - R_i i_{L,m}(t)$$

(5.8)

Since $z_m(t + \Delta t)$ is a function of $i_{L,m}(t)$, the peak currents cannot be solved directly from the maximum current $i_{L,m}(t)$ due to their nonlinear relationship with OCV.

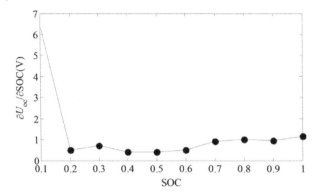

Figure 5.1 Derivative of OCV as a function of SOC.

A Taylor-series expansion is used to linearize the equation and obtain the approximate peak currents. The following is the Taylor-series expansion at the specific OCV:

$$U_{oc,m}(z_m(t + \Delta t)) = U_{oc,m}\left(z_m(t) - i_{L,m}(t)\frac{\eta_i \Delta t}{C_{max}}\right)$$

$$= U_{oc,m}(z_m(t)) - i_{L,m}(t)\frac{\eta_i \Delta t}{C_{max}}\frac{\partial U_{oc}(z)}{\partial z}\bigg|_{z=z_m(t)} + R_n\left(z_m(t), i_{L,m}(t)\frac{\eta_i \Delta t}{C_{max}}\right) \quad (5.9)$$

Considering the first-order residual $R_n(\bullet)$ and the SOC variation per sampling interval is too small to affect the OCV at the next Δt, $R_n(\bullet)$ is viewed as zero. The derivative of the OCV, $\partial U_{oc}(z)/\partial z$ can be calculated based on the OCV–SOC relationship, and its result is shown in Figure 5.1 [3]. Then, the peak currents can be solved as follows:

$$\begin{cases} i_{min,m}^{chg,volt} = \dfrac{U_{oc}(z_k(t)) - U_{t,max}}{\dfrac{\eta_i \Delta t}{C_{max}}\dfrac{\partial U_{oc}(z)}{\partial z}\bigg|_{z_m(t)} + R_{chg}} \\[6ex] i_{max,m}^{dis,volt} = \dfrac{U_{oc}(z_m(t)) - U_{t,min}}{\dfrac{\eta_i \Delta t}{C_{max}}\dfrac{\partial U_{oc}(z)}{\partial z}\bigg|_{z_m(t)} + R_{dis}} \end{cases} \quad (5.10)$$

where $i_{min,m}^{chg,volt}$ and $i_{max,m}^{dis,volt}$ are the minimum charge current and the maximum discharge current of the cell m with the voltage-limited method, respectively. $\partial U_{oc}(z)/\partial z$ is positive for most batteries within the entire SOC range, so the values computed by Eq. (5.10) are smaller than those computed by Eq. (5.2) at the same SOC. Compared with the HPPC method, the voltage-limited method allows larger Δt and takes the SOC variation into account. However, as the relaxation effect has been ignored in the R_{int} model, this method is not capable of taking polarization voltage into account in the SOP calculation.

5.2.4 MCD Method

Compared with other equivalent circuit models (ECMs), the Thevenin model [5], which has one resistance–capacitor (RC) network, is often used to simulate dynamic characteristics of batteries [4, 6–9]. To calculate peak currents based on this model, the traditional

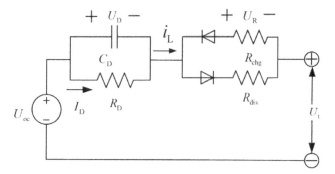

Figure 5.2 Improved Thevenin model.

ohmic resistance R_i is replaced by R_{chg} during charging and R_{dis} during discharging, which leads to the improved Thevenin model as shown in Figure 5.2.

The improved Thevenin model is mainly composed of three parts: the open circuit voltage U_{oc}, internal resistances, and an equivalent capacitance. The internal resistances include the ohmic resistances R_{chg} and R_{dis} (collectively referred to as R_i) and the polarization resistance R_D. The ideal capacitor C_D is utilized to depict transient response. U_D and U_R stand for the voltage across C_D and ohmic resistance, respectively, while I_D is the current flowing into the C_D branch. The state space representation of the improved Thevenin model is described as follows:

$$\begin{cases} U_t = U_{oc} - U_D - R_i i_L \begin{cases} R_i = R_{dis}(\text{during discharge process}) \\ R_i = R_{chg}(\text{during charge process}) \end{cases} \\ \dot{U}_D = \frac{i_L}{C_D} - \frac{U_D}{C_D R_D} \end{cases} \quad (5.11)$$

where i_L is the load current with a positive value at discharge and a negative value at charge, and U_t is the terminal voltage. Discretization of Eq. (5.11) gives

$$U_{D,m}(t + \Delta t) = \exp(-\Delta t/\tau)U_{D,m}(t) + R_D[1 - \exp(-\Delta t/\tau)]i_{L,m}(t) \quad (5.12)$$

where the time constant of polarization is defined as $\tau = R_D C_D$. Similar to the voltage-limited method which is based on the R_{int} model, the peak currents based on the improved Thevenin model can be expressed as follows:

$$\begin{cases} i_{min,m}^{chg,MCD} = \dfrac{U_{oc}(z_m(t)) - U_{D,m}(t)\exp(-\Delta t/\tau) - U_{t,max}}{\dfrac{\eta_i \Delta t}{C_{max}} \dfrac{\partial U_{oc}(z)}{\partial z}\bigg|_{z_m(t)} + R_D[1 - \exp(-\Delta t/\tau)] + R_{chg}} \\[4ex] i_{max,m}^{dis,MCD} = \dfrac{U_{oc}(z_m(t)) - U_{D,m}(t)\exp(-\Delta t/\tau) - U_{t,min}}{\dfrac{\eta_i \Delta t}{C_{max}} \dfrac{\partial U_{oc}(z)}{\partial z}\bigg|_{z_m(t)} + R_D[1 - \exp(-\Delta t/\tau)] + R_{dis}} \end{cases} \quad (5.13)$$

where $i_{min,m}^{chg,MCD}$ and $i_{max,m}^{dis,MCD}$ are the minimum charge current and maximum discharge current of the cell m with the MCD method, respectively. $\partial U_{oc}(z)/\partial z$ is positive within the entire SOC range, and the polarization voltage $U_D > 0$ when the battery is discharged and $U_D < 0$ when the battery is charged, so the values computed by Eq. (5.13) are smaller than those from Eq. (5.10) for the same ohmic resistance values. Since $\partial U_{oc}(z)/\partial z$ is not constant within the entire SOC range, especially at the two extreme ends (low and high range of the SOC), the peak power estimate based on the MCD method can provide accurate and reliable peak current calculations to improve the safety of battery systems in EVs.

As the current limits of cells have been considered, the discharge/charge currents of a battery system within all limits are expressed as

$$\begin{cases} i^{chg}_{min} = \max(i_{min}, \max_m i^{chg,SOC}_{min,m}, \max_m i^{chg,MCD}_{min,m}) \\ i^{dis}_{max} = \min(i_{max}, \min_m i^{dis,SOC}_{max,m}, \min_m i^{dis,MCD}_{max,m}) \end{cases} \tag{5.14}$$

where i_{max} and i_{min} are the maximum discharge current and the minimum charge current, respectively. The peak powers of a battery system are the sum of the powers from all the paralleled cells. They are written as

$$\begin{cases} P^{chg}_{min} = n_p \sum_{m=1}^{n_s} i^{chg}_{min} U_{t,m}(t + \Delta t) \approx \\ \quad n_p \sum_{m=1}^{n_s} i^{chg}_{min}(U_{oc}\left(z_m(t) - i^{chg}_{min}\frac{\eta_i \Delta t}{C_{max}}\right) - \exp(-\Delta t/\tau) \times U_{D,m}(t) \\ \qquad -\{R_i + R_D[1 - \exp(-\Delta t/\tau)]\}i^{chg}_{min} \\ P^{dis}_{max} = n_p \sum_{k=1}^{n_s} i^{dis}_{max} U_{t,m}(t + \Delta t) \approx \\ \quad n_p \sum_{m=1}^{n_s} i^{dis}_{max}(U_{oc}\left(z_m(t) - i^{dis}_{max}\frac{\eta_i \Delta t}{C_{max}}\right) - \exp(-\Delta t/\tau) \times U_{D,m}(t) \\ \qquad -\{R_i + R_D[1 - \exp(-\Delta t/\tau)]\}i^{dis}_{max} \end{cases} \tag{5.15}$$

where the power of each paralleled cell is equal to the product of the maximum allowable current and the estimated voltage.

It should be noted that the maximum and minimum cell power limits can also be considered in the SOP calculation if they are provided. U_{oc}, C_{max}, $U_{t,max}$, $U_{t,min}$, z_{max}, z_{min}, $I_{L,max}$, $I_{L,min}$, R_{chg}, and R_{dis} in all equations for this MCD method may be affected by temperature, SOC, and other factors related to operating conditions of battery systems.

5.2.5 Case Study

A battery system with only one pack ($n_s = 1$, $n_p = 9$) is used to demonstrate the aforementioned four peak power estimation methods. The nominal capacity of the pack is 27 Ah and the design limits on the SOC, terminal voltage, and current are shown in Table 5.1.

In order to compare the performances of the HPPC method, SOC-limited method, voltage-limited method, and MCD method, they are all used to calculate the peak currents under a federal urban driving schedule (FUDS). The current, voltage, and SOC profiles of this FUDS testing data are shown in Figure 2.10.

Figure 5.3 shows the real-time peak current estimation using the four methods. Figure 5.3a shows the peak discharge currents calculated by the HPPC method and

Table 5.1 Design limits for a LiB pack.

Parameters	Lower limit	Upper limit
SOC (%)	35	85
U_t (V)	2.65	3.65
i_L (A)	−135	270

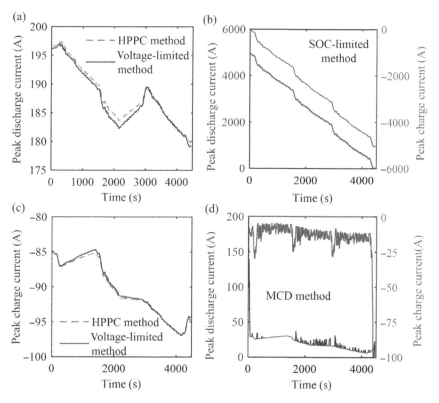

Figure 5.3 Peak currents estimation results: (a) peak discharge current with HPPC method and voltage-limited method. (b) Peak charge and discharge currents with SOC-limited method. (c) Peak charge current with HPPC method and voltage-limited method. (d) Peak charge and discharge currents with MCD method.

voltage-limited method. It can be observed that the peak discharge current obtained by the HPPC method is higher than that of the voltage-limited method. This can be attributed to the fact that $\partial U_{oc}(z)/\partial z > 0$ over the entire SOC range is taken into account in the voltage-limited method. However, the HPPC and voltage-limited methods are both based on the R_{int} model which cannot simulate dynamic characteristics of cells. By comparing Figure 5.3a with Figure 5.3d, we can observe that during discharging, polarization voltage $U_D > 0$, so it is apparent that the current obtained by the MCD method based on Thevenin model is lower than that from the HPPC and voltage-limited methods. Figure 5.3b shows that the SOC-limited method gives a higher peak current estimation. However, the peak discharge current declines and the peak charge current increases along with the decrease of the SOC in the operating condition. This feature can help to control peak currents in a reasonable range, which cannot be achieved by the HPPC and voltage-limited methods. The SOC-limited method is therefore usually combined with other methods to estimate the peak power of the battery pack.

Figure 5.3c shows that the absolute value of peak charge current calculated by the HPPC method is higher than that calculated by the voltage-limited method, which is also ascribed to the influence of $\partial U_{oc}(z)/\partial z$. The charge current calculated by the MCD method based on Thevenin model is shown in Figure 5.3d. It can be observed that if the SOC value reaches the upper limit, the peak charge current will be very small while the peak discharge current will be very large. On the contrary, if the SOC value reaches the lower limit, the peak discharge current will be very small while the peak charge current will be very large. The MCD method can optimize the peak power of the battery. Moreover, if the battery is discharged with high currents, the calculated peak discharge capability reduces significantly while the peak charge capability increases significantly. This complies with the actual working characteristics of batteries and underlying optimization control.

The peak power estimated by the MCD method is compared with that calculated by the widely used HPPC method, and the results are plotted in Figure 5.4 [2]. Figure 5.4a shows that the peak discharge power calculated by the HPPC method is higher than that

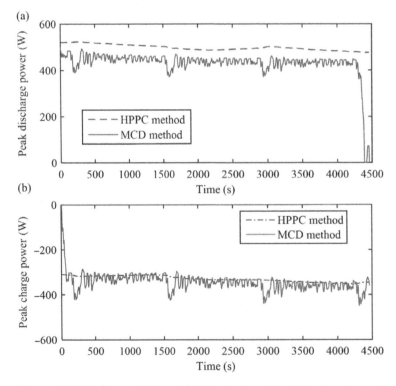

Figure 5.4 Peak power real-time results of estimation: (a) peak discharge power; (b) peak charge power.

calculated by the MCD method because the HPPC method only considers the battery voltage constraints and neglects the constraints of batter current and SOC. The HPPC method may provide an optimistic estimation of the charge power at high and low SOCs and its estimation results will inevitably be larger as $\partial U_{oc}(z)/\partial z$ is ignored, leading to over-discharging of the battery pack. In contrast, a notable feature of the MCD method is that it takes all the constraints of the battery pack into account when making real-time estimation and allows less power to be discharged.

The HPPC and MCD methods are also compared with respect to charge power, as shown in Figure 5.4b [2]. Because the SOC constraint and the dynamic performance of the battery are ignored, the HPPC method is prone to over-estimate the charging power. When there is high discharge current, the battery allows high charge power without being adjusted very rapidly. Nevertheless, the real-time peak power estimation by the MCD method changes rapidly due to the polarization effect. Therefore, the MCD method could provide a more satisfactory result. It should be noted that the change of the constraints and battery model parameters can significantly affect the SOP estimation.

5.3 Continuous SOP Estimation Method

In the previous section, the SOC-limited method, the voltage-limited method and the MCD method were introduced to calculate the SOP (or peak current) for a single sampling interval. The peak current is affected by the period of the sampling interval Δt which depends on the sampling frequency of BMSs. In principle, the SOC-limited method, the voltage-limited method, and the MCD method can be extended to calculate the SOP for multiple sampling intervals, leading to continuous SOP estimation [10]. Since the MCD method is more complex and provides more accurate estimation for continuous SOP than the SOC-limited method and the voltage-limited method, a detailed discussion is only given for the MCD method in the following section.

5.3.1 Continuous Peak Current Estimation

The Thevenin model is used to estimate the continuous peak current. In the model, the OCV is formulated as a function of SOC:

$$U_{oc} = K_0 + K_1 \ln SOC + K_2 \ln(1 - SOC) \tag{5.16}$$

Its state space equation is expressed as

$$\begin{cases} \mathbf{X}_{k+1} = \mathbf{A}\mathbf{X}_k + \mathbf{B}\mathbf{u}_k + \omega_k \\ \mathbf{Y}_{k+1} = \mathbf{C}\mathbf{X}_{k+1} + \mathbf{D}\mathbf{u}_k + v_k \end{cases} \tag{5.17}$$

where the matrices are expressed as:

$$\mathbf{X}_k = \begin{pmatrix} U_{D,k} \\ z_k \end{pmatrix}, \mathbf{Y}_k = U_{t,k}, \mathbf{u}_k = I_{L,k}, \mathbf{A} = \begin{pmatrix} \exp\left(\dfrac{-\Delta t}{\tau}\right) & 0 \\ 0 & 1 \end{pmatrix}, \mathbf{D} = [-R_i]$$

$$\mathbf{B} = \begin{pmatrix} R_D\left(1 - \exp\left(\dfrac{-\Delta t}{\tau}\right)\right) \\ \dfrac{\eta_i \Delta t}{C_a} \end{pmatrix}, \mathbf{C} = \dfrac{\partial U_t}{\partial X}\bigg|_{X=\hat{X}_{k+1}} = \begin{bmatrix} -1 & \dfrac{dU_{oc}(z)}{dz}\bigg|_{z=z_{k+1}^-} \end{bmatrix} \tag{5.18}$$

The input of the system is assumed to be constant between the kth sampling time and the $(k+L)$th sampling time, namely $\mathbf{u}_{k+L} = \mathbf{u}_k$. As the input is constant from time t_k to t_{k+L}, we obtain

$$\mathbf{X}_{k+L} = \mathbf{A}^L \mathbf{X}_k + \left(\sum_{j=0}^{L-1} \mathbf{A}^{L-1-j} \mathbf{B} \right) \mathbf{u}_k \tag{5.19}$$

Similar to Eq. (5.13), to find the minimum and maximum charging currents $I_{\min,L}^{chg}$ and $I_{\max,L}^{dis}$, the output of the battery system should meet the following two equations

$$U_{oc}(z_{k+L}) - U_{D,k}\left(\exp\left(\frac{-\Delta t}{\tau}\right)\right)^L$$
$$- I_{\min,L}^{chg}\left(R_i + R_D\left(1 - \exp\left(\frac{-\Delta t}{\tau}\right)\right)\sum_{j=0}^{L-1}\left(\exp\left(\frac{-\Delta t}{\tau}\right)\right)^{L-1-j}\right) - U_{t,\min} = 0 \tag{5.20}$$

$$U_{oc}(z_{k+L}) - U_{D,k}\left(\exp\left(\frac{-\Delta t}{\tau}\right)\right)^L$$
$$- I_{\max,L}^{dis,EP}\left(R_i + R_D\left(1 - \exp\left(\frac{-\Delta t}{\tau}\right)\right)\sum_{j=0}^{L-1}\left(\exp\left(\frac{-\Delta t}{\tau}\right)\right)^{L-1-j}\right) - U_{t,\max} = 0 \tag{5.21}$$

Then, the peak current at the period from the kth to the $(k+L)$th sampling time can be expressed by

$$\left\{ \begin{array}{l} I_{\max,L}^{dis} = \dfrac{U_{oc}(z_k) - U_{D,k}\left(\exp\left(\frac{-\Delta t}{\tau}\right)\right)^L - U_{t,\min}}{\dfrac{\eta_i L \Delta t}{C_{\max}}\dfrac{dU_{oc}(z)}{dz}\Big|_{z_k} + R_D\left(1 - \exp\left(\frac{-\Delta t}{\tau}\right)\right)\sum_{j=0}^{L-1}\left(\exp\left(\frac{-\Delta t}{\tau}\right)\right)^{L-1-j} + R_i} \\[6ex] I_{\min,L}^{chg} = \dfrac{U_{oc}(z_k) - U_{D,k}\left(\exp\left(\frac{-\Delta t}{\tau}\right)\right)^L - U_{t,\max}}{\dfrac{\eta_i L \Delta t}{C_{\max}}\dfrac{dU_{oc}(z)}{dz}\Big|_{z_k} + R_D\left(1 - \exp\left(\frac{-\Delta t}{\tau}\right)\right)\sum_{j=0}^{L-1}\left(\exp\left(\frac{-\Delta t}{\tau}\right)\right)^{L-1-j} + R_i} \end{array} \right. \tag{5.22}$$

Note that if the L is set as one, Eq. (5.22) is reduced to Eq. (5.13).

5.3.2 Continuous SOP Estimation

To ensure the safety of a battery system and meet its performance requirements, when the SOC is close to the lower limit $z_{\min}$ the maximum discharge current should be limited and the charge current can be maximized, otherwise the battery may be at risk of over-discharge. On the other hand, when the SOC is close to the upper SOC limit $z_{\max}$ the maximum charge current should be limited and the discharge current can be maximized, otherwise the battery will suffer from overcharge. This is a key principle of the peak current estimation using the SOC-limited method, which can be expressed by

$$
\begin{cases}
I_{\text{min},L}^{\text{chg,SOC}} = \dfrac{z_k - z_{\max}}{\eta_i L \Delta t / C_{\max}} \\
I_{\text{max},L}^{\text{dis,SOC}} = \dfrac{z_k - z_{\min}}{\eta_i L \Delta t / C_{\max}}
\end{cases}
\tag{5.23}
$$

where $I_{\text{min},L}^{\text{chg,SOC}}$ and $I_{\text{max},L}^{\text{dis,SOC}}$ are the minimum charge current and maximum discharge current, respectively, for the period of $L \times \Delta t$ sampling intervals. Once the design limits for the currents are given, the peak currents can be calculated as:

$$
\begin{cases}
I_{\max}^{\text{dis}} = \min(I_{\max}, I_{\text{max},L}^{\text{dis,SOC}}, I_{\text{max},L}^{\text{dis}}) \\
I_{\min}^{\text{chg}} = \max(I_{\min}, I_{\text{min},L}^{\text{chg,SOC}}, I_{\text{min},L}^{\text{chg}})
\end{cases}
\tag{5.24}
$$

where $I_{\max}$ and $I_{\min}$ are the design limits for the maximum allowable discharge current and minimum allowable charge current, respectively. $I_{\min}^{\text{chg}}$ and $I_{\max}^{\text{dis}}$ are the minimum charge current and maximum discharge current, respectively. Then, the peak power capability estimation can be formulated as:

$$
\begin{cases}
P_{\min}^{\text{chg}} = \max(P_{\min}, U_{t,k+L}^{\text{chg}} I_{\min}^{\text{chg}}) \\
P_{\max}^{\text{dis}} = \min(P_{\max}, U_{t,k+L}^{\text{dis}} I_{\max}^{\text{dis}})
\end{cases}
\tag{5.25}
$$

where $P_{\max}$ and $P_{\min}$ are the battery's power design limits; $P_{\max}$ is the peak discharge power and $P_{\min}$ is the peak charge power. $U_{t,k+L}$ can be calculated as:

$$
U_{t,k+L}^{\text{chg}} = U_{\text{oc}}(z_{k+L}) - U_{D,k}\left(\exp\left(\frac{-\Delta t}{\tau}\right)\right)^L
$$

$$
- I_{\min}^{\text{chg}}\left(R_i + R_D\left(1 - \exp\left(\frac{-\Delta t}{\tau}\right)\right)\sum_{j=0}^{L-1}\left(\exp\left(\frac{-\Delta t}{\tau}\right)\right)^{L-1-j}\right)
$$

$$
U_{t,k+L}^{\text{dis}} = U_{\text{oc}}(z_{k+L}) - U_{D,k}\left(\exp\left(\frac{-\Delta t}{\tau}\right)\right)^L
$$

$$
- I_{\max}^{\text{dis}}\left(R_i + R_D\left(1 - \exp\left(\frac{-\Delta t}{\tau}\right)\right)\sum_{j=0}^{L-1}\left(\exp\left(\frac{-\Delta t}{\tau}\right)\right)^{L-1-j}\right)
\tag{5.26}
$$

5.3.3   Influences of Battery States and Parameters on SOP Estimation

In the previous sections, it is assumed that battery SOC and model parameters are known and battery capacity is constant when we calculate the SOP. In this section, we introduce the joint estimation method which can estimate the SOP, SOC, and model parameters (including battery capacity) so that the influences of battery SOC and model parameters can be taken into the account in the SOP estimation. The AEKF is used to estimate the SOC while RLS is used to estimate the model parameters.

5.3.3.1   Uncertainty of SOC

The uncertainty of the SOC is taken into account in the SOP estimation by introducing the joint estimation for the SOC and SOP based on the AEKF. Figure 5.5 shows the flowchart of the joint estimation of SOC and SOP using the AEKF-based method, which is also known as a six-step joint estimation method.

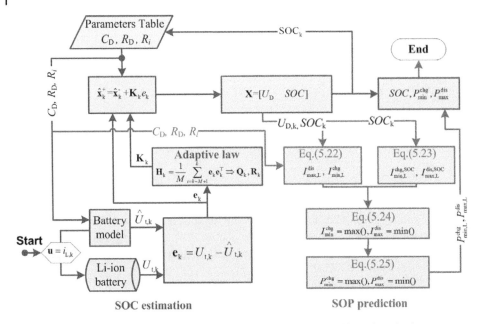

Figure 5.5 Flowchart of joint estimation of SOC and SOP using the AEKF-based method.

The corresponding calculation steps are summarized as follows:

Step 1: Initialization. Given a random initial **X** matrix, the model parameters can be identified, which are ready for the AEKF-based online estimation approach. This initialization will be finished after the program code is loaded onto the equipment.

Step 2: Start. When the desired load current or power is given, the charge/discharge current is loaded to the battery and the battery model simultaneously.

Step 3: Correction and SOC estimation. The terminal voltage error between the observer and the experimental data is adaptively reduced by updating the gain of the AEKF, where the updated gain is used to compensate for the state estimation error. The SOC estimation is then fed back to update the parameters of the battery model for the SOC estimation at the next sampling time. An accurate estimation for SOC and U_D will be obtained.

Step 4: Dynamic peak current estimation. We calculate the peak currents using Thevenin model and the SOC limits, which are performed by Eqs. (5.22) and (5.23).

Step 5: The MCD method. When taking the design limits of current, voltage, and power into consideration, we can achieve the peak power estimation with Eqs. (5.24) and (5.26).

Step 6: End. Finally, we can achieve the joint estimation for the SOC and SOP.

5.3.3.2 Case Study

The AEKF-based SOC and SOP joint estimation approach is applied to a 3.7 V/35 Ah lithium-ion battery (LiB) at different initial SOCs under the urban dynamometer driving schedule (UDDS) test. The design limits of this LiB on the SOC, terminal voltage,

Table 5.2 Design limits on SOC, terminal voltage, current, and power for continuous peak power estimation.

Parameter	Minimum value	Maximum value
SOC (%)	15	90
U_t (V)	3.0	4.25
i_L (A)	−175	350
P (W)	−750	1500

Table 5.3 Identified model parameters.

SOC (%)	C_D (F)	R_D (mΩ)	R_i (mΩ)	τ (s)
60	29 029	1.142	1.914	33.16
70	26 484	1.335	1.925	35.36
80	29 250	1.045	1.936	30.57

current, and power for peak power capability estimation are shown in Table 5.2; they are normally given by battery and vehicle manufacturers. The design limit of the SOC may be different for different control strategies in EVs.

The peak current and peak power design limits are restricted for continuous power estimation. It should be noted that with increase in duration, the peak currents and powers will become higher. The model parameters of this LiB are identified and listed in Table 5.3.

The AEKF-based joint estimation of the SOC and SOP is conducted on this LiB when the battery is fully charged at the SOC of 100%. We show the estimation results by setting correct and incorrect SOCs.

The estimation results obtained by setting a correct initial SOC of 100% are shown in Figure 5.6. It can be seen that the estimated terminal voltage can mimic the terminal voltage obtained from the experiment well with the voltage error less than 50 mV.

The results indicate that the AEKF-based approach can accurately estimate terminal voltage by timely adjusting the KF gain matrix K_k according to the voltage error. As a result, the estimated SOC tracks its reference well with the SOC error within the range of −1% and +1%, proving the high accuracy of the SOC estimation.

The peak power capabilities at the time intervals of 15, 30, and 60 seconds are predicted by using the AEKF-based SOC and SOP joint estimation method, and the results are shown in Figure 5.7. It can be observed that the peak current and power decrease with the increase of the sampling interval, proving that the instantaneous peak power estimation methods are not appropriate for continuous peak power prediction. Otherwise, the battery will have the risk of being overcharged or over discharged.

The SOC estimation results obtained by setting the incorrect initial SOCs of 95% and 50% are shown in Figure 5.8. Under such conditions, the AEKF-based SOC and SOP joint estimation is performed under UDDS. It can be seen that the SOC estimation results are almost the same as their reference SOCs after convergence. The zoomed-in

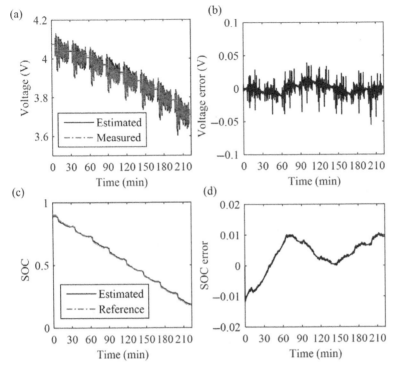

Figure 5.6 AEKF-based SOC estimation results by setting a correct initial SOC: (a) terminal voltage; (b) voltage error; (c) SOC; (d) SOC error.

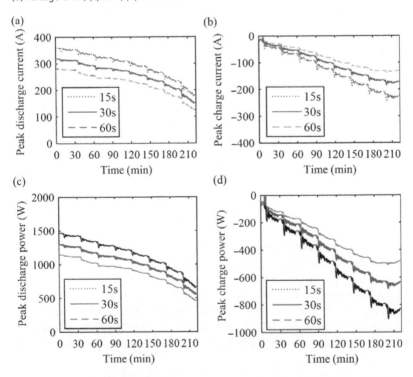

Figure 5.7 AEKF-based SOP estimation by setting a correct initial SOC: (a) peak discharge current; (b) peak charge current; (c) peak discharge power; (d) peak charge power.

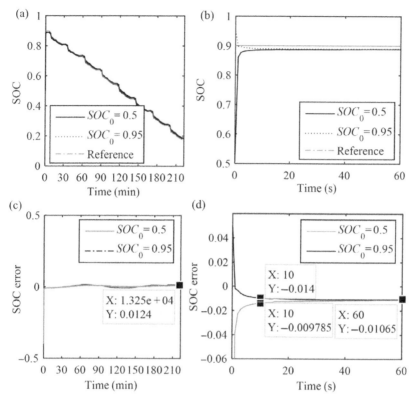

Figure 5.8 SOC estimation obtained by the AEKF-based joint approach: (a) SOC estimation with initial SOCs of 95% and 50%; (b) zoomed-in figure from 0 to 60 seconds taken from (a); (c) SOC estimation error with initial SOCs of 95% and 50%; (d) zoomed-in figure from 0 to 60 seconds taken from (c).

figure in Figure 5.8b shows the detailed performances of the AEKF-based SOC and SOP estimation approach within 60 seconds. It indicates that the AEKF-based joint estimation method can ensure the convergence of the SOC after 10 sampling intervals. The SOC estimation error is shown in Figure 5.8c and the zoomed-in figure in Figure 5.8d shows the detailed SOC estimation error within 60 seconds. It further indicates that the SOC estimation can converge to the reference SOC within 10 sampling intervals and after convergence the estimation error can be reduced to 2%.

With the accurate SOC estimation, the AEKF-based SOC and SOP estimation approach can further be used to estimate peak current and power. The peak discharge current estimation is used as an example. Figure 5.9a shows the estimation of peak discharge current with one true initial SOC and two erroneous initial SOCs for a sampling period of 30 seconds and Figure 5.9b shows the corresponding zoomed-in figure within 60 seconds. It can be seen that the estimated peak currents with two erroneous initial SOCs are almost the same as the estimated peak current with the true initial SOC (Figure 5.9a) and the estimation results can converge to the stable value within 10 sampling intervals (Figure 5.9b). Therefore, the AEKF-based joint estimation approach can effectively estimate the SOC and SOP accurately regardless of large initial SOC errors.

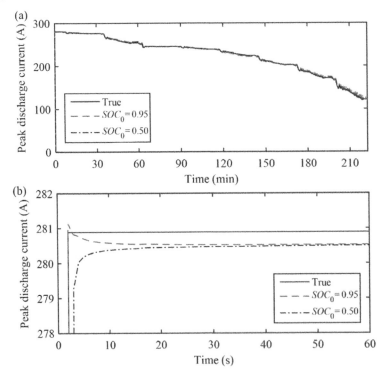

Figure 5.9 Peak current estimation results under two inaccurate initial SOCs: (a) estimated peak discharge current and true value; (b) zoomed-in figure taken from (a).

5.3.3.3 Uncertainty of Model Parameters

As battery model parameters are sensitive to SOC and have a large effect on SOC and SOP estimation, an online SOC and SOP joint estimation method is designed to improve the estimation accuracy and deal with the uncertainty of model parameters by combining the RLS and AEKF [11]. The flowchart of the joint estimation of the SOC and SOP is presented in Figure 5.10 [11]. In this method, the RLS method is used to identify the model parameters in real-time so that the accurate battery model can be obtained to further improve the estimation accuracy of the SOC and SOP. There are four steps to jointly estimate the SOC and SOP with model parameters identified in real-time:

Step 1: Onboard data measurement. After the discharge and charge current profiles are loaded on the LiB, the real-time measurement of battery current and voltage are executed by the voltage and current sensors at each sampling interval.

Step 2: Online parameter identification. With the measured current and voltage, the RLS method is used to identify the battery model parameters during each sampling interval.

Step 3: Adaptive SOC estimation. The identified parameters and the voltage errors between the model and experimental data are transferred to the AEKF-based SOC estimator. The AEKF algorithm adjusts the SOC to update the OCV and then corrects the model output voltage to minimize the voltage error until the voltage error converges to the predefined value. At the same time, the SOC will converge to its reference trajectory, resulting in accurate SOC estimation.

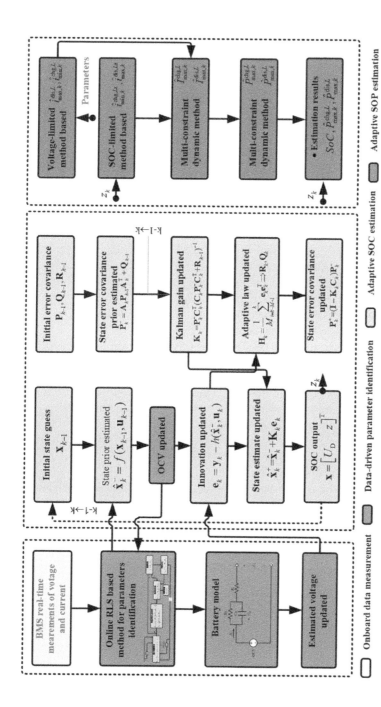

Figure 5.10 Flowchart of SOC and SOP joint estimation based on RLS and AEKF algorithms.

Step 4: Adaptive SOP estimation. With the real-time model parameter identification and the accurate SOC estimation at each sampling interval, the charge and discharge current capability and SOP are calculated by the MCD method. This will lead to a reliable multi-state joint estimation of SOC and SOP.

It is obvious that the other types of SOC estimation algorithms can also work with the continuous peak power estimation method to achieve joint estimation of the SOC and SOP.

5.3.3.4 Case Study

The LiB cells at different aging levels are used to elaborate this method. To calculate the SOP, the design limits on voltage, current, and power of the LiB cells have been preset by battery manufacturers. The SOC design limits of the tested cells are also given to meet the requirements of the control strategies for EVs. An example of these limits is given in Table 5.4, where the limits of current and power capabilities are restricted for the SOP calculation for 30 seconds of continuous output.

Figure 5.11 shows the estimated results of the voltage and SOC for the LiBs under the Dynamic Stress Test (DST) with the initial SOC set to the true value of 100%. Figure 5.11a compares the estimated and measured terminal voltages, and their differences (or voltage estimation errors) are given in Figure 5.11b. Similarly, the estimated and reference SOCs, and their differences (or SOC estimation errors) are shown in Figure 5.11c and Figure 5.11d, respectively. The maximum voltage error is less than 15 mV and the maximum SOC error is less than 0.1%.

Figure 5.12 shows the estimation results of the terminal voltage and SOC with an erroneous initial SOC. Figure 5.12a compares the estimated and measured terminal voltage, and their differences (or voltage estimation errors) are given in Figure 5.12b. Figure 5.12c compares the estimated and reference SOCs, and their differences (or SOC estimation errors) are given in Figure 5.12d. It can be observed that the estimated voltage converges to the measured voltage and the SOC converges to the reference SOC within a few seconds. After convergence, the maximum voltage error is about 10 mV and the maximum SOC error is less than 1%.

Figure 5.13 shows the continuous estimation of peak charge and discharge currents for the LiB (30 seconds), where the initial SOC is set to the true value of 100%. Figure 5.13a,b shows the charge and discharge currents, respectively, estimated by the MCD method, the SOC-limited method, and the voltage-limited method. Figure 5.13c,d shows the charge and discharge currents, respectively, estimated only by the MCD method.

When the SOC is close to its upper design limit, the SOC-based method can potentially make the battery avoid safety hazards from overcharge effectively. When the SOC

Table 5.4 Design limits for power capability estimation (30 seconds).

Parameters	Max. value	Min. value
SOC (z_{max}, z_{min}) (%)	100	20
U_t $(U_{t,max}, U_{t,min})$ (V)	4.05	3.0
I_L (I_{max}, I_{min}) (A)	320	−160
P (P_{max}, P_{min}) (W)	1000	−450

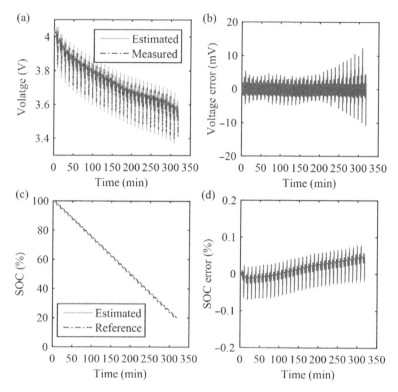

Figure 5.11 Estimation results with correct initial SOC: (a) voltage; (b) voltage error; (c) SOC; (d) SOC error.

is close to its lower design limit, the SOC-based method can also potentially make the battery avoid safety hazards from over-discharge. As a result, when the SOC is close to its design limits, the estimated current capability may fluctuate, as shown in Figure 5.13c,d.

Figure 5.14 shows the estimated SOP. It can be seen that the estimated peak charge power gradually decreases with the increase of the SOC, and the estimated peak discharge power gradually decreases with the decrease of the SOC.

To illustrate the relationship between the peak current and the continuous time which this peak current will apply, the peak charge and discharge current is estimated at different continuous times as shown in Figure 5.15. These results are obtained by the method described in Figure 5.10.

Figure 5.15 shows that the current capability estimation markedly increases with the decrease of the continuous sampling intervals. The discharge current capability for the continuous sampling interval of 1 second is similar to the level of a short circuit current which is much larger than that of 30, 60, and 120 seconds. The charge current capability decreases greatly with the increase of the continuous time. Thus, the continuous charge current should not be set with high value or it will easily exceed the allowable charge currents which may lead to potential safety hazards from overcharge. The current and power capability estimation at the correct and erroneous initial SOCs for the continuous time of 30 seconds is conducted and the results are plotted for comparison in Figure 5.16.

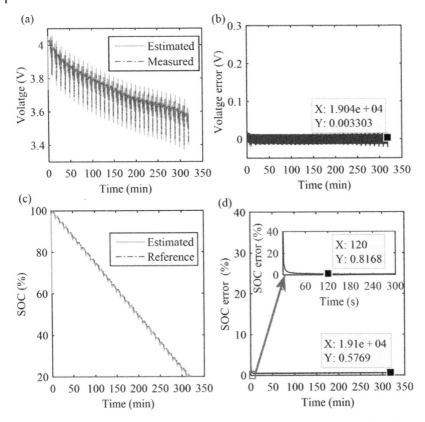

Figure 5.12 Estimation results with erroneous initial SOC of 60%: (a) voltage; (b) voltage error; (c) SOC; (d) SOC error.

Combining with the results shown in Figure 5.12, we find that the estimated current and power capability with different initial SOCs are basically the same after the convergence within the limited sampling intervals, indicating that this method enjoys a good robust performance against the erroneous settings of initial SOCs.

5.3.3.5 Uncertainty of SOH

Battery aging affects charge and discharge capability. The SOP will be estimated with the DST testing data of the LiB cells at different aging levels as discussed in Chapter 2. The results are shown in Figure 5.17. It should be noted that the time length of the DST test is different due to the different available capacities of the three cells.

In Figure 5.17, it is found that the estimated SOP of the battery has a direct relationship to its degradation state. This is due to the following two reasons: (i) the internal resistance usually increases with the degradation of battery health status, which will lead to the reduction of charge and discharge capacity under the same conditions. (ii) The OCV range will also be narrowed down, namely, the OCV value at 100% SOC will decrease with the increase of the aging levels and the OCV value at 0% SOC will increase with the increase of the aging level. According to Eq. (5.13), higher internal resistance and lower

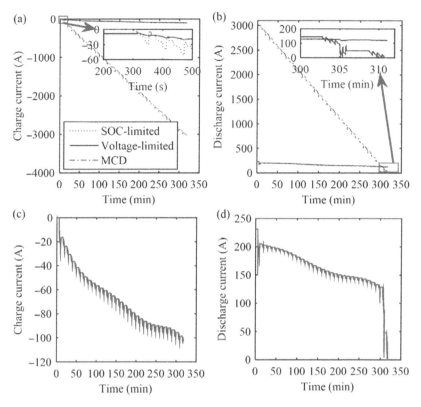

Figure 5.13 Estimation results with accurate initial SOC (30 seconds): (a) charge current comparison of three methods; (b) discharge current comparison of three methods; (c) charge current estimated by MCD method; (d) discharge current estimated by MCD method.

OCV value at 100% SOC will result in a reduction of the peak discharge capability of the battery. It can be seen from Figure 5.17b that the peak discharge power of the LiB cell with a capacity of 25.7 Ah is much smaller than those of the other two LiB cells with capacities of 29.5 and 31.6 Ah.

For the peak charge current, lower OCV value at 100% SOC has a more significant effect on the estimation of the peak power capability than higher internal resistance. The absolute value of the estimated peak charge power of the LiB with a capacity of 25.7 Ah is larger than that of the other two cells when their SOCs are near 100%. This phenomenon can also be explained from Eq. (5.13). In this equation, the battery terminal voltage $U_{t,max}$ is the charge voltage, when the OCV U_{oc} becomes smaller the absolute value of the difference between the terminal voltage and OCV increases, leading to large charge current and thus large charge power estimation result. This estimation result shows that the peak charge power of the battery near 100% SOC increases with aging process.

Therefore, the joint SOC and SOP estimation method can provide accurate and robust SOC and SOP estimation results, and more importantly, the joint estimation of SOC and capacity needs to be combined with SOP estimation to achieve joint estimation of SOC, state of health (SOH), and SOP.

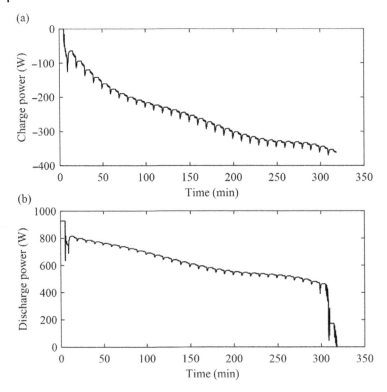

Figure 5.14 SOP estimation with accurate initial SOC (30 seconds): (a) charge power; (b) discharge power.

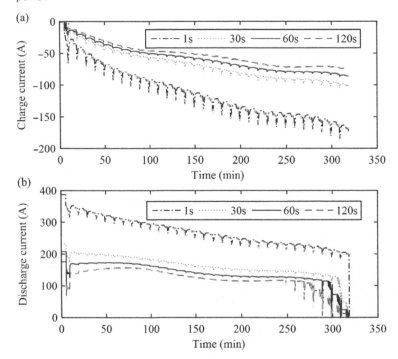

Figure 5.15 Peak current estimation for different continuous times: (a) charge current; (b) discharge current.

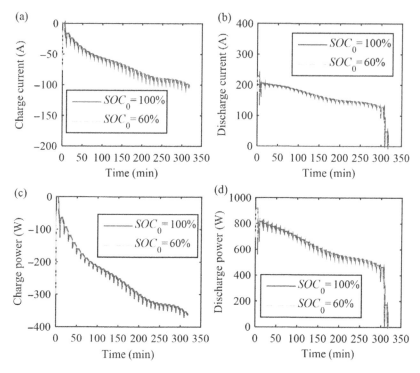

Figure 5.16 Current and power capability estimation results with correct and erroneous initial SOCs (30 seconds): (a) charge current; (b) discharge current; (c) charge power; (d) discharge power.

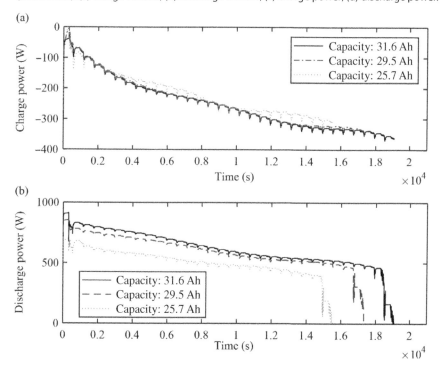

Figure 5.17 Power capability estimation for the cells with different aging levels (30 seconds): (a) charge power; (b) discharge power.

5.4 Summary

The SOP is closely related to the safety and efficiency of BMSs in EVs. In general, the SOP estimation methods can be divided into instantaneous and continuous estimation methods based on the length of the continuous time of estimation. For instantaneous SOP estimation methods, four methods are introduced, namely a HPPC method, a SOC-limited method, a voltage-limited method, and a MCD method.

The SOC-limited method, the voltage-limited method, and the MCD are then extended to the continuous SOP estimation. A detailed explanation is only given for the MCD method. In the MCD method, the Thevenin model is used to capture the polarization effect of a battery. The influences of SOC, SOH and model parameters on the SOP estimation are also discussed by using the joint SOC and SOP estimation based on the AEKF and RLS. The joint SOC and SOP estimation based on the MCD method is more reliable for on-board EV applications.

References

1 Belt, J.R. (2010). *Battery Test Manual for Plug-in Hybrid Electric Vehicles*. Idaho National Laboratory.

2 Sun, F., Xiong, R., He, H. et al. (2012). Model-based dynamic multi-parameter method for peak power estimation of lithium-ion batteries. *Applied Energy* 96: 378–386.

3 Plett, G.L. (2004). High-performance battery-pack power estimation using a dynamic cell model. *IEEE Transactions on Vehicular Technology* 53 (5): 1586–1593.

4 He, Y., Liu, W., and Koch, B.J. (2010). Battery algorithm verification and development using hardware-in-the-loop testing. *Journal of Power Sources* 195 (9): 2969–2974.

5 Salameh, Z.M., Casacca, M.A., and Lynch, W.A. (1992). A mathematical model for lead-acid batteries. *IEEE Transactions on Energy Conversion* 7 (1): 93–98.

6 He, H., Xiong, R., and Fan, J. (2011). Evaluation of lithium-ion battery equivalent circuit models for state of charge estimation by an experimental approach. *Energies* 4 (4): 582–598.

7 Roscher, M.A. and Sauer, D.U. (2011). Dynamic electric behavior and open-circuit-voltage modeling of $LiFePO_4$-based lithium ion secondary batteries. *Journal of Power Sources* 196 (1): 331–336.

8 Kim, J., Lee, S., and Cho, B.H. (2011). Discrimination of Li-ion batteries based on Hamming network using discharging–charging voltage pattern recognition for improved state-of-charge estimation. *Journal of Power Sources* 196 (4): 2227–2240.

9 He, H., Xiong, R., and Chang, Y.-H. (2010). Dynamic modeling and simulation on a hybrid power system for electric vehicle applications. *Energies* 3 (11): 1821–1830.

10 Xiong, R., He, H., Sun, F. et al. (2013). Model-based state of charge and peak power capability joint estimation of lithium-ion battery in plug-in hybrid electric vehicles. *Journal of Power Sources* 229: 159–169.

11 Xiong, R., Sun, F., He, H., and Nguyen, T.D. (2013). A data-driven adaptive state of charge and power capability joint estimator of lithium-ion polymer battery used in electric vehicles. *Energy* 63: 295–308.

6

Battery Charging

6.1 Background

Battery charging in electric vehicles (EVs) is similar to refueling a tank in internal combustion engine vehicles (ICEVs). Battery charging involves many issues such as charging scheme, charging power transfer, charging facility, and charging algorithm. Among them, the charging algorithm is the key to battery charging, which has a direct impact on energy usage and the cycle life of battery systems in EVs. This chapter first provides a brief introduction to the charging scheme, charging power transfer, and charging facility, and then there is a detailed discussion of charging algorithms.

Many charging schemes have been developed to recharge battery systems in EVs. They are mainly divided into four categories [1]. The first category is normal charge. It usually takes 5–8 hours to fully charge battery systems. It facilitates load leveling when performing charging overnight in a residential garage or car park without causing extra burden to power systems. It generally adopts a single-phase AC, 110–240 V, 13–20 A, and 2–4 kW. The second category is opportunity charge. It allows charging of battery systems in a public car park many times whenever possible as long as battery systems are not fully charged, and extends EV driving range without causing significant burden to power systems. It typically uses a three-phase AC, 110–240 V, 32–80 A, and 8–20 kW. The third category is fast charge. It charges battery systems to 80% of their normal capacity within 20–30 minutes at the dedicated fast charging station due to the high power demand. Such high power may cause safety issues and extra burden to power systems, especially during daytime peak hours. It generally adopts DC, 200–450 V, 80–200 A, and 36–90 kW. The fourth category is battery swap. It mechanically swaps discharged and charged battery systems within five minutes. As a result, the discharged battery systems can be scheduled to conduct charging during off-peak hours. The standardization of all EV battery systems is the main concern of this charging scheme.

Charging power transfer can be performed through wire or wireless. The use of conductive wires for power transfer to charge battery systems in EVs is simple, with low cost and high efficiency. The disadvantages are lack of convenience and the risk of electrocution. In contrast, the use of wireless techniques for power transfer to charge battery systems is complicated, with high cost and low efficiency. Due to the absence of conductive wires, the risk of possible electrocution during the charging process is totally eliminated, which enables EVs to outperform ICEVs in terms of convenience and safety. There are two main groups of wireless power transfer (WPT) techniques: far-field and near-field. Far-field WPT uses microwave radiation or laser to transfer

Advanced Battery Management Technologies for Electric Vehicles, First Edition. Rui Xiong and Weixiang Shen.
© 2019 John Wiley & Sons Ltd. Published 2019 by John Wiley & Sons Ltd.

high power over long distances. It requires complicated tracking strategies and large antennas, which are impractical for EVs. Near-field WPT employs electric field or magnetic field for short-range to mid-range power transfer including capacitive power transfer (CPT) via electric field and inductive power transfer (IPT) via magnetic field. Among the near-field WPT techniques, IPT has been identified as the most viable technology to charge battery systems for EVs [2]. The working principle of IPT is based on magnetic coupling between the primary and secondary coils of a high-frequency transformer. The primary coil is installed in a charger coupler while the secondary coil is embedded in a vehicle inlet. In the side of the charger coupler, the main AC supply with a frequency of 50 or 60 Hz is rectified and converted into a high-frequency AC power of about 80 kHz for power transfer to the vehicle side by induction. In the vehicle side, the high-frequency AC power is converted into DC power for battery charging. The IPT operates over a wide frequency range and is readily scaled up to meet various power levels for EV charging.

The charging facility can be implemented as plugless chargers by using IPT. The park-and-charge (PAC) and move-and-charge (MAC) systems are two of the most convenient charging facilities to be developed compared with normal conductive chargers in EVs. In PAC, the primary coil is installed in the floor of a garage or parking lot and the secondary coil is installed in the vehicle [3]. The system is very easy to use, and the charging process is automatic once the driver has parked the EV correctly. Due to a large air-gap or clearance between the primary and secondary coils, the magnetic resonant coupling technique is usually adopted, where two coils that have the same resonant frequency wirelessly transfer power efficiently with high power density while dissipating relatively little energy in nonresonant objects such as vehicle bodies or drivers. In MAC, an array of power transmitters is embedded beneath a roadway known as a charging zone or lane while a receiver is mounted on the bottom of the EVs. MAC can conveniently charge EVs at the charging zone during driving, extending EV driving distances. However, many challenges need to be overcome before the mass application of PAC and MAC to EVs [3].

The charging algorithm defines a charging profile to charge a battery cell in terms of current and voltage over time and a condition to terminate the charging process. Generally, the constant current (CC) charging profile is used to charge nickel–cadmium (NiCd)/nickel–metal hydride (NiMH) batteries, whereas the constant voltage (CV) charging profile is used to charge lead–acid batteries [4]. For the termination of a charging process, voltage drop (ΔV) is normally incorporated with the CC to stop charging NiCd/NiMH batteries, whereas the cut-off current is normally incorporated with the CV to stop charging lead–acid batteries. There are many other termination techniques, such as timer, temperature cut-off, temperature change rate (dT/dt), cut-off voltage for charge, cut-off voltage for discharge and voltage change rate (dV/dt).

Today, Li-ion batteries are used in almost all commercial EVs [5]. Various charging algorithms have been developed for Li-ion batteries [1, 6]. They can be divided into the constant current/constant voltage (CC/CV) and variants of the CC/CV, multistep constant current (MSCC), two-step constant current/constant voltage (TSCC/CV), constant voltage constant current constant voltage (CVCC/CV), constant current pulse charging (CC-PC), and constant voltage pulse charging (CV-PC). Recently, there have been two new developments for consideration in charging Li-ion batteries. One is to find the optimal charging current profile (CCP) for Li-ion batteries based on the minimization of cost function such as energy losses during the entire charging process and the other is to develop a Li-ion battery with an extremely high charging and discharging capability. This Li-ion battery uses the spinel lithium titanate as the negative electrode to replace the graphite negative electrode in the lithium titanate oxide (LTO) battery. The LTO battery can be charged extremely fast simply by the CC charging profile. By way of an introduction to each of the charging algorithms, the basic terms for the evaluation of charging performances and characterizations are first defined in the following section.

6.2   Basic Terms for Evaluating Charging Performances

Various basic terms are defined to evaluate charging performances [4]. These terms are summarized below.

6.2.1   Cell and Pack

A single cell is a complete battery with two current leads and separate compartment holding electrodes, separator, and electrolyte. A battery pack consists of many single cells connected in series and/or parallel with battery balancing and battery management systems. More than one battery pack is needed for EVs.

6.2.2   Nominal Ampere-Hour Capacity

The ampere-hour (Ah) capacity of a battery is the total charge that can be discharged from the fully charged battery. The nominal capacity (C_n) is the rated Ah capacity of a fully charged battery under the conditions specified by a manufacturer.

6.2.3   C-rate

C-rate is used to represent a charging or discharging rate. It equals the ratio of the nominal capacity to the duration of charge or discharge in hours. For a 2.3 Ah Li-ion battery, 1C charging or discharging rate represents 2.3 A. Likely, 0.1C and 10C represent 0.23 A and 23 A, respectively.

6.2.4   Cut-off Voltage for Discharge or Charge

Cut-off voltage for discharge ($V_{cut-off}^{dis}$) is the minimum allowable voltage and can be interpreted as the "empty" of a battery at the discharging rate and temperature defined by a manufacturer. It is normally used to stop the discharging process in the period of CC discharging. Cut-off voltage for charge ($V_{cut-off}^{chg}$) is the maximum allowable voltage defined by a manufacturer. It is a temperature-dependent parameter. It is normally used to stop the charging process in the period of CC charging.

6.2.5   Cut-off Current

Cut-off current ($I_{cut-off}^{chg}$) is the predefined current which is normally very small (e.g. 0.01C for Li-ion batteries) and used to stop the charging process during the period of CV charging.

6.2.6   State of Charge

State of charge (SOC) is defined as the ratio of the remaining capacity (C_r) to the nominal capacity (C_n) of a battery

$$Z = C_r/C_n \tag{6.1}$$

or

$$Z(t) = Z(t_0) - \frac{1}{C_n} \int_{t_0}^{t} \eta I(\tau)d\tau \tag{6.2}$$

where $Z(t_0)$ is the initial SOC, η is the charging or discharging efficiency, and $I(\tau)$ is the current which is negative for charging or positive for discharging. The SOC is a critical parameter for battery management in EVs and the key to the healthy and safe operation of batteries.

6.2.7   State of Health

State of health (SOH) is defined as the ratio of the capacity of the aged battery (C_{aged}) to the nominal capacity of the new battery. It is an important parameter to indicate the degree of performance degradation of a battery and to estimate the remaining lifetime of the battery.

$$SOH = C_{aged}/C_n \tag{6.3}$$

6.2.8   Cycle Life

Cycle life is the number of charging and discharging cycles when the battery capacity fails to 80% of nominal capacity of the new battery in EVs. The actual cycle life is affected by the charging and discharging rates and other conditions such as temperature.

6.2.9   Charge Acceptance

Charge acceptance describes the share of the current that can actually be accepted by a battery and could be retrieved by subsequently discharging. It is determined by the balance between the current which converts electrical energy into chemical energy stored in the battery and the current which contributes to the internal losses of the battery. Charge acceptance is not a constant parameter, but depends largely on the SOC. It decreases with the increase of the SOC.

6.2.10   Ampere-Hour Efficiency

Ampere-hour efficiency (η_{Ah}) is the ratio of the discharged capacity (C_{dis}) to the charged capacity of a battery (C_{chg}) under the specific charging and discharging conditions

$$\eta_{Ah} = C_{dis}/C_{chg} = \int_{0}^{t_{dis}} i_{dis}(t)dt / \int_{0}^{t_{chg}} i_{chg}(t)dt \tag{6.4}$$

The Ah efficiency for Li-ion batteries is very high and close to 100% [7].

6.2.11 Ampere-Hour Charging Factor

The ampere-hour charging factor (α_{Ah}) is the ratio of the charged capacity (C_{chg}) to the discharged capacity (C_{dis}) under the specific charging and discharging conditions

$$\alpha_{Ah} = C_{chg}/C_{dis} \tag{6.5}$$

If the battery is fully charged, the discharged capacity is the nominal capacity, namely $C_{dis} = C_n$, then

$$\alpha_{Ah} = C_{chg}/C_n \tag{6.6}$$

6.2.12 Energy Efficiency

Energy efficiency (η_{en}) is the ratio of the discharged energy (E_{dis}) to the charged energy of a battery (E_{chg}) under the specific charging and discharging conditions

$$\eta_{en} = E_{dis}/E_{chg} = \int_0^{t_{dis}} v_{dis} i_{dis}(t) dt / \int_0^{t_{chg}} v_{chg} i_{chg}(t) dt \tag{6.7}$$

Since the battery terminal voltage during the charging process is much higher than that during the discharging process, the energy efficiency for Li-ion batteries is much lower than the Ah efficiency and its value is generally varied from 85 to 95% in the wide range of temperatures and charging rates [7].

6.2.13 Watt-Hour Charging Factor

The watt-hour charging factor (α_{wh}) is the ratio of the charged energy of a battery (E_{chg}) to the discharged energy (E_{dis}) under the specific charging and discharging conditions

$$\alpha_{wh} = E_{chg}/E_{dis} \tag{6.8}$$

If the battery is fully charged, then

$$\alpha_{wh} = E_{chg}/(C_n \times V_n) \tag{6.9}$$

where V_n is the normal voltage of the battery.

6.2.14 Trickle Charging

Trickle charging (TC) means continuous charge at low charging current ($I_{trickle}^{chg}$) approximately equivalent to the internal losses of a battery and suitable to maintain the battery in the fully charged state. The current required for TC amounts to 0.1C or 0.01C depending on battery types.

6.3 Charging Algorithms for Li-Ion Batteries

When the terminal voltage of a Li-ion battery reaches cut-off voltage for discharge or the SOC of a Li-ion battery reaches 20% or lower, the Li-ion battery must stop discharging and needs to be recharged. When the battery is charged, a charging algorithm must be adapted to charging characteristics of Li-ion batteries. In addition to the charging characteristics, the battery charging process is also affected by the amplitudes of charging current or voltage or a combination of both as well as the environmental temperature surrounding the battery. The key criterion to the charging algorithms is to recharge the batteries to their full capacity without causing extended overcharge or excessive temperature. Based on this criterion, many charging algorithms for Li-ion batteries have been developed to meet different charging schemes. The following sections provide the detailed discussions of these charging algorithms.

6.3.1 Constant Current and Constant Voltage Charging

The CC/CV charging is the most common charging algorithm for Li-ion batteries. It starts with a high charging rate in the CC mode at the initial stage and ends at the appropriately regulated voltage in the CV mode which limits overcharge. Since it is the most popular charging algorithm for Li-ion batteries, it has become a benchmark with which the performances of the newly developed charging algorithms for Li-ion batteries are compared. In the CC/CV charging algorithm, a regulated charging current ($I_{cc/cv}$) is first applied to charge the batteries until the battery voltage rises to a predefined voltage ($V_{cc/cv}$). Then, such a charging voltage ($V_{cc/cv}$) is maintained which leads the charging current to decrease exponentially. The charging process stops at the time (t_{end}) when the charging current reaches a predefined cut-off current ($I_{cut\text{-}off}^{chg}$). Figure 6.1 shows the typical terminal voltage and current profiles when the CC/CV charging is applied to charge a Li-ion battery.

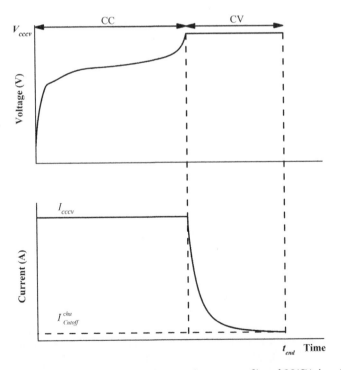

Figure 6.1 Typical terminal voltage and current profiles of CC/CV charging for Li-ion batteries.

The selection of the charging current in the CC step depends on the types of Li-ion batteries and their charge acceptance. It can be generally chosen to be 0.5C to 3.2C for the charging time in the range of 0.5–2.5 hours [8]. The selection of the charging voltage in the CV steps also depends on the types of Li-ion batteries. For the four most commercially used Li-ion batteries, the typical charging voltage settings are 4.2 V for lithium cobalt oxide (LCO), $LiCoO_2$, lithium manganese oxide (LMO), $LiMn_2O_4$, and lithium nickel manganese cobalt oxide (NMC), $LiNiMnCoO_2$, and 3.6 V for lithium iron phosphate (LFP), $LiFe_2PO_4$ [9]. Figure 6.2 shows the CC/CV charging voltage profiles at a CC of 1C and a CV of 4.2 V for a LMO battery and a CV of 3.6 V for a LFP battery.

To implement a CC/CV charger, precautionary measures are taken to protect the batteries. Figure 6.3 shows the charging process when the CC/CV is used to charge Li-ion batteries [10]. It shows that the charging process of the CC/CV consists of three steps. First, the battery initial conditions, such as temperature and open circuit voltage (OCV), are checked if they are in the normal range. If the OCV is less than the cut-off voltage for discharge ($V^{dis}_{cut-off}$) (e.g. 2.5 V for LMO), the battery is charged by TC with a small current of 0.1C until the battery voltage (V_b) rises to $V^{dis}_{cut-off}$. Secondly, once the battery voltage exceeds $V^{dis}_{cut-off}$, the CC mode starts to charge the battery at the regulated charging current (I_{cccv}). Thirdly, when the battery voltage reaches the cut-off voltage for charge (e.g. $V^{chg}_{cut-off} = 4.2$ V for LMO), the charging process switches to the CV mode, the battery is charged at the regulated voltage ($V_{cc/cv}$) and the charging current is reduced exponentially. The charging period is terminated by either cut-off current ($I^{chg}_{cut-off}$) which is typically 0.01C for Li-ion batteries or the maximum charging time ($I^{chg}_{cut-off}$).

Depending on the charging current in the CC mode, the total charging time is varied. In general, the lower the charging current of the CC mode, the higher the charging efficiency and the longer the charging time and the battery life. Three sensors are required to measure and monitor battery voltage, current, and temperature, making sure that they are all within the maximum allowable values. The main advantage of the CC/CV charging algorithm is that no model information is required to charge the battery. Furthermore, the CC control and the CV control can be easily realized with very simple

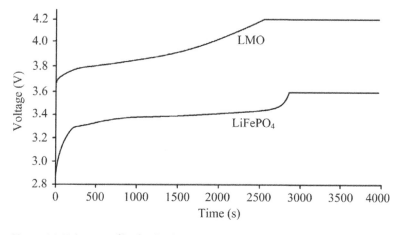

Figure 6.2 Voltage profiles for the CC/CV charging algorithm with a CC of 1C at a voltage of 3.6 V for a LFP battery and at a voltage of 4.2 V for a LMO battery.

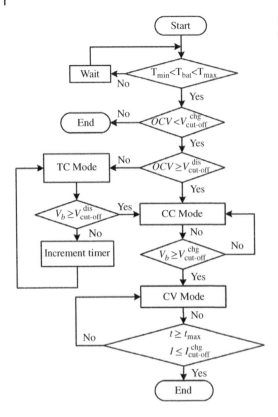

Figure 6.3 Charging process of a CC/CV charger.

integrated circuits (ICs) or a microcontroller at low cost. The main drawback is that there is a long "tail" in the CV charging period which prolongs the charging time.

Recently, many variants of the CC/CV charging algorithm have been developed to charge Li-ion batteries. One of these variants uses a simple double-loop (DL) controller to achieve the charging profile similar to the conventional CC/CV charging profile (DL-CC/CV), as shown in Figure 6.4 [11]. The DL-CC/CV is implemented by using the negative and positive feedbacks of the battery voltage ($V_B(s)$) to the battery reference voltage ($V_{Br}(s)$) and the output of the PID controller. The large error ($e_B(s) = V_{Br}(s) - V_B(s)$) creates a similar charging process to the CC charging while the minimization of $e_B(s)$ creates a charging process equivalent to the CV charging. Since only the battery terminal voltage is required to measure, the need for a current sensor is eliminated. As a result, the DP-CC/CV achieves similar performances to the conventional CC/CV with the requirement of only voltage and temperature sensors at low cost.

$V_{Br}(S)$ $V_{Bs}(S)$ $V_B(S)$

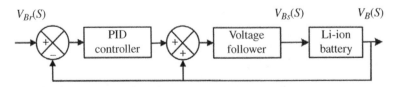

Figure 6.4 Block diagram of a double-loop control charger.

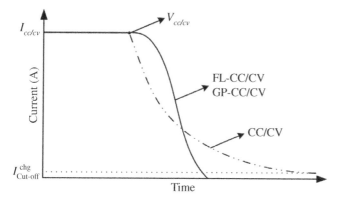

Figure 6.5 Charging current profiles of FL-CC/CV, GP-CC/CV and CC/CV.

The other two variants use advanced control schemes to realize the charging profile similar to that of the conventional CC/CV charging algorithm. The advanced control schemes are the fuzzy logic (FL) control [12] and the gray-predicted (GP) control [13]. For the former, a FL-controlled active SOC controller is used to realize CC/CV (FL-CC/CV). For the latter, a GP technique is used to realize CC/CV (GP-CC/CV). The essence of these two charging algorithms is to replace the regulated terminal voltage in the charging process with the OCV (e.g. $V_{oc} = 4.2\,V$) so that the appropriate charging current in the CV mode is dynamically determined based on the SOC and the charging current in the CV charging period under the conventional CC/CV. As a result, the charging current of the CV mode is larger in the higher current part and smaller in the lower current part than the charging current in the CV mode of the conventional CC/CV, eliminating the long "tail" of the CV charging as shown in Figure 6.5. Thus, more capacity is able to be charged into the battery in the period of the CV mode and a shorter time is required to complete the CV mode. The FL-CC/CV and the GP-CC/CV have a shorter charging time and a higher charging efficiency. Due to their complexity and the requirement of high computation power, a microcontroller is more suitable for the realization of both charging algorithms.

Another variant of the CC/CV charging algorithm is implemented by applying the principle of the phase-locked loop (PLL) control to the charging process [14]. The PLL process naturally coincides with the requirement of the charging process of the CC/CV.

Figure 6.6 shows the block diagram of the PLL-based CC/CV charging algorithm (PLL-CC/CV). The working principle of the PLL-CC/CV can be explained in three steps. First, the battery voltage (V_b) is measured and sent to a difference amplifier. The output

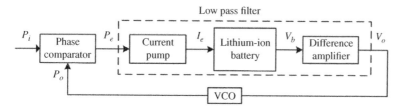

Figure 6.6 Block diagram of PLL-CC/CV.

of the amplifier (V_o) is conditioned to adapt the input voltage scale of the voltage control oscillator (VCO). Secondly, the VCO generates a feedback phase (P_o) that represents the battery voltage and sends to a phase comparator. The phase comparator then compares the input reference phase (P_i) to produce the phase error (P_e). Finally, this phase error is sent to the current pump to produce a suitable current to charge lithium-ion batteries.

In this PLL-CC/CV arrangement, the auto-tracking process, i.e. the frequency-tracking process, corresponds to the regulated current charging or bulk charging which is similar to the CC mode of the CC/CV. The auto-locking process from the phase-tracking to the phase-locked state corresponds to the variable current charging and float charging which is similar to the regulated voltage charging in the CV mode of the CC/CV. Figure 6.7 shows the flow chart for the charging process of the PLL-CC/CV.

Later, a current-pump-based PLL-CC/CV (CPLL-CC/CV) is proposed to improve the charging process of the PLL-CC/CV [15]. The charging process of the CPLL-CC/CV consists of the regulated current charging (CC mode) which remains the same as that of the PLL-CC/CV, but the variable current charging and float charging of the PLL-CC/CV under the regulated voltage are modified to the pulsed current charging and the pulsed current float charging, respectively, as shown in Figure 6.8. As a result, the internal pressure of the battery charged by a pulse current is smaller than a CC due to the insertion

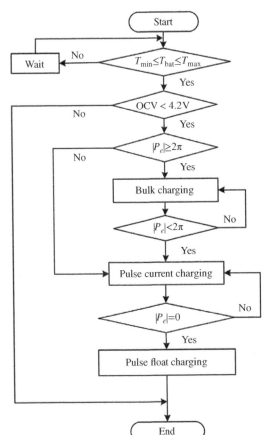

Figure 6.7 Flow chart for the charging process of PLL-CC/CV.

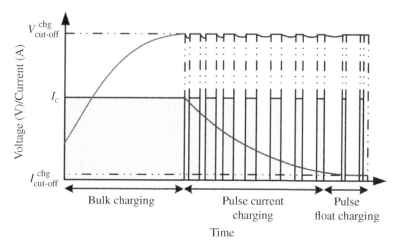

Figure 6.8 Charging profile of CPPL-CC/CV.

of a short rest period. Comparing with the PLL-CC/CV, the CPLL-CC/CV has high charging efficiency. The total charging time is similar to that of the CC/CV. Both the PLL-CC/CV and the CPLL-CC/CV can be easily implemented by using the specific IC with the PLL function.

6.3.2 Multistep Constant Current Charging

The MSCC charging algorithm has been explored to charge Li-ion batteries. The main idea of the MSCC charging algorithm is to use the reduced current amplitude in multistep during the charging process to minimize the possibility of overcharge and to improve charging efficiency. The adoption of the reduced CC is due to the fact that the charge acceptance of the batteries is gradually decreased with the progressing recharge.

The MSCC was first proposed to charge lead–acid batteries [16]. The charging procedure mainly includes four steps. In the first step, the charging current is chosen at C/6 before reaching the gassing voltage of 2.45 V. Then, the charging current is cut-off so that the cell voltage is allowed to decay until it reaches a predefined value of 2.2 V. In the second and third steps, the charging currents are taken as 75% and 50% of that in step 1, respectively, to charge the battery up to the gassing voltage and then decay down to the predefined voltage. In the fourth step, the charging current is chosen as 25% of that in step 1 to continue charging the battery over the gassing voltage until no rise in the cell voltage is detected within 15 minutes which leads to charging termination.

Later, experiments of the MSCC with two, three, four, and six steps were investigated for lead–acid batteries [17]. It is found that the MSCC with four and six steps can further shorten the charging time with higher energy efficiency and longer cycle life than the MSCC with two steps. The MSCC is then explored to charge NiMH batteries [18]. The experiments of MSCC with one, two, and three steps are investigated to seek the optimal charging algorithms for NiMH batteries with the improvement of charging efficiency, charging time and cycle life. It is shown that the MSCC with three steps results in the shortest charging time (less than two hours) with reasonable charging efficiency and cycle life. The first two steps are controlled by the preset cut-off voltage for charge

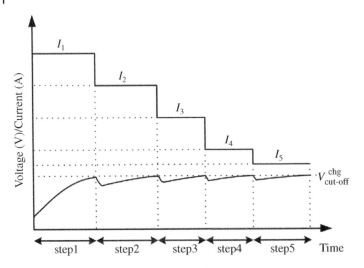

Figure 6.9 Charging profile of MSCC with five steps.

($V_{\text{cut-off}}^{\text{chg}}$) and the last step is regulated by the changing rate of the battery temperature; if dT/dt is equal to a predefined value (α_T) (e.g. $\alpha_T = 0.13\,^\circ\text{C}\,\text{min}^{-1}$), the charging process will be terminated.

Recently, the MSCC has been explored to charge Li-ion batteries with five steps, as illustrated in Figure 6.9. The amplitude of charging current in each step must be appropriately determined to charge the battery.

Four approaches are proposed to optimize charging currents in the MSCC charging algorithms. The first approach uses a FL controller to determine a charging current [19], where the inputs of the FL controller are the temperature and the change of the temperature, and the output of the FL controller is the charging current. The second approach uses the consecutive orthogonal array (or Taguchi method) to search an optimal CCP [20, 21]. The third approach applies an ant colony system to optimize a CCP [22]. The fourth approach applies an integer linear programming method to search an optimal CCP [23].

The MSCC with four steps has also been explored to charge a Li-ion battery recently [24]. In this approach, the Taguchi method is employed to search an optimal charging current pattern and an adaptive switching gain sliding mode observer is employed to estimate the SOC which controls and terminates the charging process. The total charging period is equally divided into four steps, where each step has the range of the SOC equal to 25%. In each step, the Li-ion battery is charged by a pre-set current. During the charging process, the SOC is estimated in real-time. When the estimated SOC reaches the predetermined SOCs, such as 25%, 50%, and 75%, the charging process will be shifted to the next step and a new pre-set charging current will be applied to charge the battery accordingly. The charging process will continue until the estimated SOC reaches 100%, namely the battery at this state is considered as fully charged and the charging process terminates. Figure 6.10 shows the CCP for MSCC with four steps. This charging algorithm has been used to charge a Li-ion battery at different SOCs of 0%, 17%, 37%, 67%, and 84%, respectively, and their experimental results are shown in Figure 6.11. It shows that this charging algorithm can charge Li-ion batteries with any SOCs, where

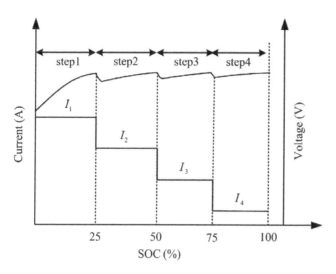

Figure 6.10 Charging profile of MSCC with four steps.

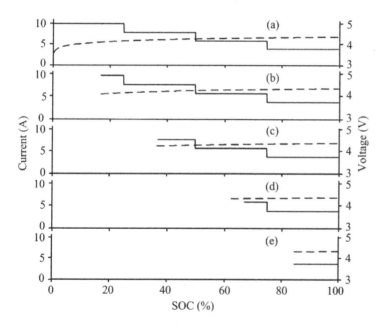

Figure 6.11 Experimental results of charging profiles of MSCC with four steps at different SOCs: (a) SOC = 0%; (b) SOC = 17%; (c) SOC = 37%; (d) SOC = 67%; (e) SOC = 84%.

the estimated SOC is used to switch the steps and stop the charging process. To be safe, the temperature and cut-off voltage for charge can also be used as the backup methods to stop the charging process together with the estimated SOC.

In summary, MSCC charging algorithms can be developed based on the flowchart shown in Figure 6.12. In this flowchart, the shaded blocks may vary from one approach to another as various methods can be chosen to optimize charging profiles in terms of

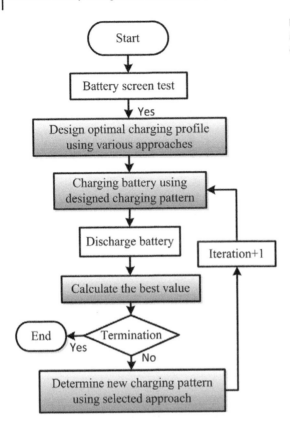

Figure 6.12 Flowchart to implement MSCC with various approaches for charging profile optimization.

the "best value" which can be the highest charging efficiency or the shortest charging time or a compromise between both; the rest of the blocks remain the same. Generally, the MSCC algorithm has higher charging speed and higher charging efficiency than the CC/CV charging algorithm. A microcontroller or a computer is required to implement this charging algorithm due to its complexity.

6.3.3 Two-Step Constant Current Constant Voltage Charging

TSCC/CV charging has been developed to charge a high power LFP battery within approximately 20 minutes [25]. The TSCC/CV charging algorithm combines the high CC, low CC and CV into one charging profile. It is developed on the basis of the evolution of internal resistance during the charging process which is divided into three steps, referred to as CC-I, CC-II, and CV-I. In the CC-I step, the battery is charged at a regulated CC (I^1_{CC-I}) (e.g. $I^1_{CC-I} = 4C$) up to the cut-off voltage for charge (e.g. $V^{chg}_{cut-off} = 3.6\,V$). The charging current of 4C is selected because it corresponds to initial low internal resistance and high charge acceptance as well as the maximum charging current recommended by the manufacturer. In the CC-II step, the battery is charged at another regulated CC (I^2_{CC-2}) (e.g. $I^2_{CC-2} = 1C$). Since the current in CC-II is much lower than that in CC-I, the battery voltage drops below 3.6 V so that the charging process can continue until the battery voltage reaches $V^{chg}_{cut-off}$ again. In the CV-I step, the battery is charged at a regulated constant voltage of 3.6 V only for a short duration of five minutes because it

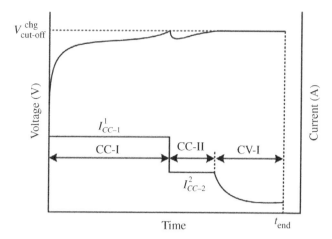

Figure 6.13 Charging profile of TSCC/CV.

avoids an acceleration of the aging process that occurs at high SOC values over longer periods of time. The extreme high charging current is used to charge the battery within the safe voltage initially, significantly shortening charging time. Figure 6.13 shows the charging profiles of the TSCC/CV charging algorithm.

6.3.4 Constant Voltage Constant Current Constant Voltage Charging

CVCC/CV has been designed to charge a Li-ion battery. It is also called boost charging due to initial high charging voltage [26]. The CVCC/CV is required to fully discharge a battery before charging. It consists of three charging steps, including the regulated high CV, regulated CC and regulated low CV. For the CVCC/CV, the initial high charging voltage (V_b^{max}) is set to 4.3 V which is 0.1 V higher than the normal cut-off voltage for charge (e.g. $V_{cut-off}^{cha} = 4.2$ V). The battery is charged at V_b^{max} in the boost charging period (t_b) (e.g. $t_b = 5$ minutes) and the charged capacity can be restored around 30% of the nominal capacity of the battery. It shows that a significant amount of charge has been stored in the battery within a relatively short period t_b. If this period is extended to 10 minutes, about 60% of the nominal capacity of the battery can be charged into the battery. After this period, the charging algorithm is then switched to the conventional CC/CV. Figure 6.14 shows the charging profile of the CVCC/CV.

Due to the initial high charging voltage, the CVCC/CV can charge the battery faster than the standard CC/CV. However, the discharging circuits are required to fully discharge the battery before charging. This will increase the number of components and cost. Discharging a battery before charging also makes this charging algorithm inefficient.

6.3.5 Pulse Charging

Pulse charging has been considered as a promising candidate for a fast charging algorithm for lead–acid, NiCd/NiMH and Li-ion batteries. It employs a pulse current or voltage to charge the battery periodically with the insertion of a rest period into a pulse

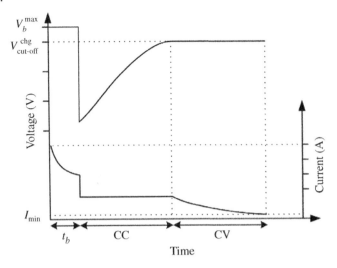

Figure 6.14 Charging profile of CVCC/CV.

interval. The rest period allows the electrolyte to distribute more evenly which leads to an efficient charging process. The effects of pulse charge on Li-ion batteries are evaluated using an electronic network model [27]. Simulation results provide some insight into the effect of the pulse on internal processes, such as diffusion, migration, electrochemical reactions, and heat generation. An experimental approach is also used to investigate the effect of pulse charge on the cycle life of a Li-ion battery [28]. Many pulse charging algorithms have been developed. They can be classified as CV-PC and CC-PC algorithms.

The CV-PC is realized by keeping the amplitude of the pulse voltage constant in the entire charging process while changing the frequency of the pulse (FCV-PC) [29] or the duty cycle of the pulse (DCV-PC) [30]. The basic idea of FCV-PC is to adjust the frequency of the pulse within a certain range and observe the response of the charging current. The optimal frequency (f_{opt}) of the pulse charging is obtained when the battery receives the highest charging current (I_{opt}) at the minimum impedance. Figure 6.15 shows the flow chart of the FCV-PC for a Li-ion battery. The DCV-PC is very similar to the FCV-PC. The only difference is that the DCV-PC changes the duty cycle of the pulse, instead of the frequency of the pulse, to achieve the highest charging current at the lowest impedance. The prototypes of implementing these two charging algorithms have been made, demonstrating that the charging time is shorter than that of the conventional CC/CV [29, 30] with higher charging efficiency and longer cycle life.

On the other hand, the CC-PC is realized by keeping the amplitude of the pulse current constant in the entire charging process while monitoring the battery voltage to make sure that it is always lower than the cut-off voltage for charge. The charging profile can be varied by changing the amplitude and width of the pulse current and the relaxation period between the pulses [31, 32]. Figure 6.16 shows the charging profile of the CC-PC. A simulation study has been conducted to optimally select the charging time (t_c) and the rest time (t_r) for the given amplitude of the pulse current. The charging time is optimized when the maximum concentration is reached, and the relaxation period is determined such that it provides sufficient time to a reset concentration. As a result,

Figure 6.15 Flow chart of FCV-PC.

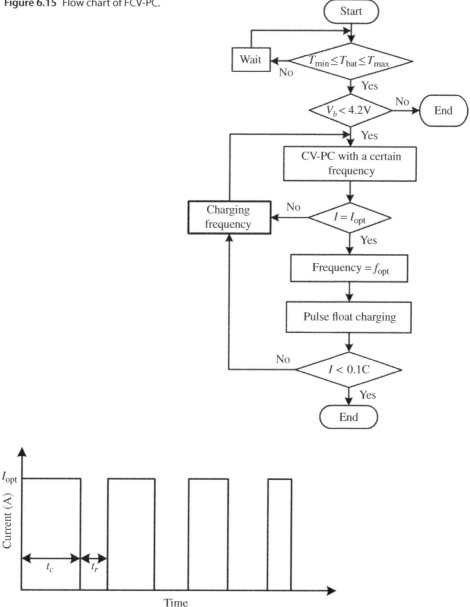

Figure 6.16 Charging profile of CC-PC.

the electrochemical reaction inside the battery neither produces heat nor causes the accumulation of pressure. Since this charging algorithm is designed based on the link between the pulse CCP and the chemical reaction process inside the battery, in principle it can charge the battery faster and more efficiently. However, it is hard to be applied in a real charger since the concentration change inside a battery is not measurable.

6.3.6 Charging Termination

Charging termination is a critical part of a charging algorithm. Ideally, the battery SOC is the best indicator to safely stop the charging process when the SOC reaches 100%. However, the SOC cannot be measured directly but has to be estimated. Due to the complexity of the SOC estimation, the estimated SOC may not be accurate and reliable. Backup methods are always required to stop the charging process by using other directly measured parameters such as the voltage, current and temperature of a battery. The following are some typical techniques to terminate the charging process for Li-ion batteries [4, 33, 34].

- *Timer.* When a pre-set timer expires, the charging process is stopped. This method is very simple but can be advantageous if the charger fails to detect a full charge state by other means. Normally, it is used as a backup together with other termination methods. It can be also used for any type of batteries.
- *Temperature cut-off.* The charging process will be stopped if the absolute temperature of a battery rises to a certain threshold value. This method is very simple and normally is used as a backup or safety measure with the support of other termination methods. It can be also used for any type of batteries.
- *Cut-off current for charge.* When the charging current reaches the preset small current, the charging process stops. This method is normally incorporated with a CV charging algorithm and is normally used for Li-ion batteries. It can be also used for lead–acid batteries.
- *Cut-off voltage for charge.* When the battery voltage reaches a threshold value, the charging process will be terminated. This method is normally incorporated with a CC charging method to stop charging Li-ion batteries. It can also be used for lead–acid batteries.

6.3.7 Comparison of Charging Algorithms for Lithium-Ion Batteries

The charging algorithms for Li-ion batteries in EVs have been discussed. The CC/CV is the conventional approach to charge Li-ion batteries. The MSCC is the advanced approach to charge Li-ion batteries. These charging algorithms vary in charging time, charging efficiency, implementation complexity and cost, impact on cycle life, and sensors required. They range from the simplest charging algorithms such as the CC/CV to the most creatively complicated charging algorithms such as MSCCs with ant colony or consecutive orthogonal array or the Taguchi method. Table 6.1 summarizes the major aspects of these charging algorithms [6]. It should be noted that all the above-discussed charging algorithms have the predefined constant amplitude of voltage or current or a combination of both during the charging process, which simplifies their implementation in a real charger.

Recently, there have been two new developments in Li-ion battery charging. One is the CCPs based on the concept of variable amplitude of current during the charging process. To obtain the optimal CCP for Li-ion batteries, the cost function of the CCP in terms of charging time [35], the trade-off between charging time and energy loss [36], charging loss [37] or energy losses [38] is established. The other is the newly developed Li-ion battery, the LTO battery ($Li_4Ti_5O_{12}$), which has an extremely fast charging and discharging capability. The underlying principle for the high charge acceptance of the

Table 6.1 Major characteristics of charging algorithms for Li-ion batteries.

Charging algorithms	Analog or digital	Charging time	Charging efficiency[a]	Implementation complexity, cost	Cycle life	Sensors required
CC	Analog	Short	Low	Low	Short	V, T
CV	Analog	Long	Medium	Low	Short	I, T
CC/CV	Analog or digital	Long	Medium	Medium	Short	V, I, T
DL-CC/CV	Analog	Long	Medium	Low	Short	V, T
FL-CC/CV	Digital	Medium	Medium	High	Medium	V, I, T
GP-CC/C	Digital	Medium	Medium	High	Medium	V, I, T
PLL-CC/CV	Analog	Long	Medium	Medium	Medium	V, I, T
CPLL-CC/CV	Analog	Long	Medium	Medium	Medium	V, I, T
MSCC	Digital	Medium	Medium	High	Long	V, I, T
TSCC/CV	Digital	Short	Medium	Medium	Long	V, I, T
CVCC/CV	Digital	Short	Low	Medium	Short	V, I, T
FCV-PC	Digital	Short	High	High	Long	V, I, T
DCV-PC	Digital	Short	High	High	Long	V, I, T
CC-PC	Analogs or Digital	Short	High	High	Medium	V, I, T

a) Charging efficiency refers to overall charging efficiency including Ah charging efficiency and energy charging efficiency.

V, voltage sensor; I, current sensor; T, temperature sensor.

LTO battery is that there are nanocrystals on the negative electrode instead of carbon, which provides a much more effective surface area. The following sections will explain these two new developments.

6.4 Optimal Charging Current Profiles for Lithium-Ion Batteries

Rather than predefining the CCPs, various optimal CCPs for Li-ion batteries have been developed. The optimal CCPs are dynamically determined with the goal of adapting battery SOC during the entire charging process. One of these optimal CCPs is developed to minimize the charging time for a Li-ion battery [35]. Different constraints on internal battery states are considered such as the over-potential in the negative electrode and the final SOC. The simulation results based on an electrochemical model suggest a decrease in charging time by 50% compared with the CC/CV. A second of these optimal CCPs has been developed to give a compromise between the charging time and the charging loss [36]. The battery equivalent circuit model (ECM) is used to analyze the influence of the maximum charging voltage, battery temperature and battery aging on charging results for both NMC and LFP batteries. The simulation results show that given the same charging time the optimal CCP has the advantage of energy-saving in terms of the battery aging over the CC/CV. A third one has been developed to charge a Li-ion battery by targeting loss minimization [37]. The optimal CCP is obtained by using an evolution algorithm to minimize the charging losses based on the variation of battery internal impedance during the charging process. The experimental results show that the loss decreases by 41.1%, 31.7%, and 14.8% for charging rates equivalent to 1C, 0.5C, and 0.25C, respectively. A fourth one has been developed to charge a Li-ion battery with optimal CCP targeting the minimization of the energy loss of a battery during the charging process [38]. This fourth process is taken as an example to explain how the optimal CCP can be obtained.

6.4.1 Energy Loss Modeling

To minimize the energy loss during the charging process, a loss model based on battery ECM is established, where the one resistance–capacitor (RC) ECM is adopted to represent a Li-ion battery as shown in Figure 6.17. This ECM consists of an OCV source (U_{ocv}), an internal resistance (R_o), a parallel branch of a capacitance (C_p), and a polarized resistance (R_p). A Li-ion polymer battery is used as the testing sample with the normal capacity of 40 Ah and normal voltage of 3.7 V in the operation range of 2.7 V ($V^{dis}_{cut-off}$) to 4.2 V ($V^{chg}_{cut-off}$). The capacity test, OCV test, hybrid pulse power characterization (HPPC) test and Urban Dynamometer Driving Schedule (UDDS) test are conducted on the selected Li-ion polymer battery to obtain the relationship between the OCV and the SOC, identify these circuit parameters and validate the accuracy of the ECM. Figure 6.18 shows the relationship between the OCV and SOC and Figure 6.19 shows the polarized resistance (R_p) and internal resistance (R_o) versus the SOC. Figure 6.20 shows the capacitance (C_p) versus the SOC.

After obtaining all the parameters, the energy loss model during the charging process can be expressed by

$$E_{loss} = \int_0^{T_{tot}} [i_p^2 R_p(Z) + i_L^2 R_o(Z)]dt \tag{6.10}$$

where T_{tot} is the total charging time. Z denotes the SOC. During the charging process, the SOC is calculated by an Ah counting method

$$Z_{T_{tot}} = Z(0) - \frac{1}{C_b} \int_0^{T_{tot}} i_L dt \tag{6.11}$$

where C_b is the normal capacity of the Li-ion polymer battery.

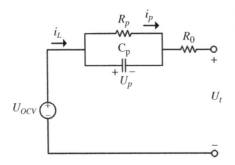

Figure 6.17 ECM of a Li-ion battery for energy loss modeling.

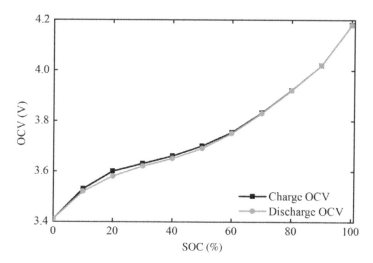

Figure 6.18 Relationship between OCV and SOC.

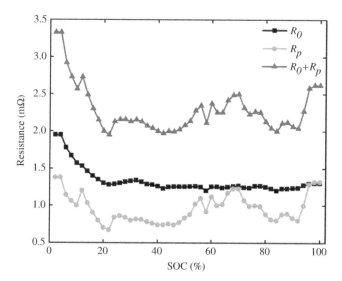

Figure 6.19 R_p and R_o versus SOC.

6.4.2 Minimization of Energy Loss

Dynamic programming (DP) is used to minimize energy loss and obtain the optimal CCP. DP formulates the minimization of energy loss during the charging process into multistep decision processes and creates the energy loss matrices to save the data that will be used in the future calculation to reduce the computation time [39].

Consider the charging process to make the battery SOC change from an initial state Z_{min} (e.g. SOC = 0%) to the final state Z_{max} (e.g. SOC = 80%) for the given charging time T_{tot}. With the given starting and ending states of the charge, many possible paths in terms of charging profiles can be chosen to charge the battery. The goal is to find an

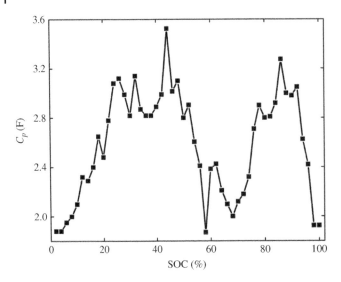

Figure 6.20 C_p versus SOC.

optimal CCP to minimize the energy loss during the charging process. To simplify the minimization of energy loss in Eq. (6.10), the charging current i_L can be assumed to be i_p. This is due to the fact that the charging current changes slowly during the charging process, under such a pseudo-steady state the capacitance C_p can be considered as an open circuit. Consequently, the energy loss from Z_k to Z_{k+1} can be expressed in a discrete form as

$$E_{\text{loss}}^i = i_{L,k}^2 [R_p(Z_k) + R_o(Z_k)](t_{k+1} - t_k) \tag{6.12}$$

The two constrains used in this DP are:

$$I_0 = [(Z_{\text{max}} - Z_{\text{min}})C_b]/T_{\text{tot}} \tag{6.13}$$

$$I_{\text{max}} = 2I_0 \tag{6.14}$$

where I_0 is the charging current and I_{max} is the maximum charging current which is defined as twice I_0. The minimum charging current I_{min} is set to zero. To solve the problem at the trade-off of acceptable accuracy and computation burden, the step current value I_{step} is defined as

$$I_{\text{step}} = I_0/50 \tag{6.15}$$

$$i_{L,k} = kI_0, k = 0, 1, 2, \ldots, 50 \tag{6.16}$$

From Eq. (6.16), there are a total of 51 choices of the current during the calculation. The DP is used to find a sequence of the optimal CCP from them. To realize the DP, the energy loss matrix with respect to different SOCs and charging current levels is constructed to build the cost-to-go matrix [40, 41], and will choose the best path to the present state as follows:

$$\text{min\_}\cos t_{Z_{k-1}} + loss_{Z_{k-1}} < \text{min\_}\cos t_{Z_k}$$

$$\text{min\_}\cos t_{Z_k} = \text{min\_}\cos t_{Z_{k-1}} + loss_{Z_{k-1}} \tag{6.17}$$

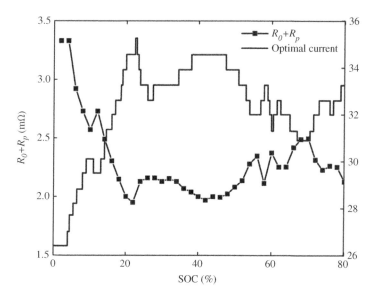

Figure 6.21 Optimal CCP versus the variation of battery resistance at different SOCs.

where $\min\_\cos t_{Z_k}$ is the minimum cost at Z_k and $loss_{Z_{k-1}}$ is the loss at the SOC of Z_{k-1}. Then, the costs of all the possible paths are compared to ensure that the minimum energy loss is obtained at the SOC of Z_k and stored as $\min\_\cos t_{Z_k}$. The above calculation is repeated until the maximum SOC and the total charging time is reached. As such, the optimal CCP can be found at the minimum energy loss. Figure 6.21 shows that the optimal CCP varies with the sum of the resistances R_o and R_p at different SOCs.

It can be seen that generally the DP allocates the charging current inversely proportional to battery resistance within the SOC range of 0–80% so that the energy loss in the charging process is minimized, namely the charging current is given a relatively higher value at a relatively lower resistance and a relatively lower value at a relatively higher resistance. The charging current varies from 26.2 to 35.2 A, but the average value of the charging current is maintained at 32 A which is the same as the charging current used in the CC mode of the traditional CC/CV charging algorithm.

6.5 Lithium Titanate Oxide Battery with Extreme Fast Charging Capability

All the charging algorithms that have been discussed in the previous sections are generally used to charge carbon-based Li-ion batteries, such as carbon-based LCO, LMO, LFP, NMC, and lithium nickel cobalt aluminum oxide (NCA), $LiNiCoAlO_2$, batteries, where the negative electrode materials are graphite/carbon related compound. Recently, the spinel LTO $Li_4Ti_5O_{12}$ has been considered as a better alternative to a carbon-based negative electrode due to its excellent electrochemical characteristics [42]. These characteristics are: (i) extremely fast Li^+ insertion and de-insertion ability leads to extremely fast charging and discharging capability and a flat operation voltage. (ii) LTO is a kind of zero strain insertion material which exhibits negligible volume change during cycling, hence showing almost no mechanical aging. (iii) the comparatively high voltage of about 1.55 V versus Li/Li^+ suppresses the formation of the solid electrolyte interphase (SEI) and lithium plating at high charging current and/or low temperature during the charging process. This allows high performance LTO batteries to be made. Take the lifetime of different LTO batteries as an example [43]. The LTO/LCO battery cell was studied experimentally. A capacity fade of 12% was observed after 5000 cycles at 100% depth of discharge (DOD). For the LTO/LMO battery cell, a capacity fade of 4% was observed after 30 000 cycles with 10C rate at 100% DOD. The LTO/LFP battery cell showed negligible capacity fade after 20 000 full cycles. The LTO/NMC had 14% capacity fade after 6000 cycles with 3C rate at 100% DOD. The LTO/NCA underwent 3400 equivalent full cycles at a symmetric 4C charge and discharge rate. The capacity fade was shown to be 1.28%/1000 cycles. All these data demonstrate that the lifetimes of the LTO-based batteries are longer than those of carbon-based Li-ion batteries. Table 6.2 shows the main characteristics of a LTO battery [44, 45].

The long cycle life, safety and fast charging and discharging capabilities of LTO batteries are advantageous features in applications where these attributes are critical. These applications include use in telecommunication base stations, UPS data centers, aerospace, sensors, renewable energy, smart grid, and EVs.

The two types of LTO batteries are used to illustrate the charging characteristics. Figure 6.22 depicts the evolution of battery SOC versus the charging time at a CC charging rate of 8C for a Toshiba SCiB 20 Ah cell. Figure 6.23 depicts the evolution of terminal voltages versus the cell charge capacity for an Altair Nano 50 Ah cell, where a CC charging rate of 1C, 5C, and 10C and a cut-off voltage for charge of 2.9 V are used in the experiments.

It can be seen from Figure 6.22 that with a charging rate of 8C the LTO battery cell can be charged to a SOC of 40% in 3 minutes, 80% in 6 minutes, and 100% (fully charged state) in 15 minutes. It can also be seen from Figure 6.22 that with a charging rate of 8C the LTO battery cell can be charged to a SOC of 20% in 1.5 minutes, 40% in 3 minutes, and approximately 80% in 6 minutes.

Table 6.2 Main characteristics of a LTO battery.

Specific energy	Approximately 30–110 $Wh\,kg^{-1}$
Energy density	As high as 177 $Wh\,l^{-1}$
Specific power	3000–5100 $W\,kg^{-1}$ (peak load)
Nominal voltage	2.3 V (1.9–2.4 V or 1.5–2.7 V, operating)
Self-discharge	2–5%/mo
Operating temperature	−35 to +55 °C
Cycle life	6000 cycles at DOC of 90%, some models > 10 000 cycles
Charging/discharging efficiency	Charging efficiency over 95%/discharging efficiency approximately 85%
Charging method/charging time	Standard CC/CV/less than 10 min (charging rate can be higher than 10C)
Lower cut-off voltage	1.5 V typical (some at 1.7 V)
Overcharge tolerance	High endurance/safety and stability

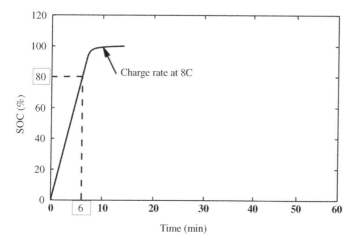

Figure 6.22 SOC versus charging time at a CC charging rate of 8C.

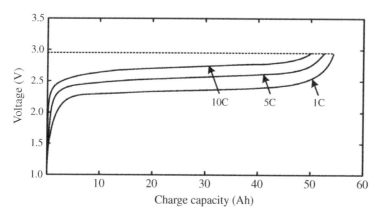

Figure 6.23 Voltages versus cell charge capacity at CC charging rates of 1C, 5C, and 10C.

6.6 Summary

In this chapter the charging scheme, charging power transfer, and charging facility are briefly introduced. Various technical terms are defined to evaluate the performances of different charging algorithms. Based on the definitions, the charging algorithms for Li-ion batteries are discussed in detail. These charging algorithms include CC/CV and variants of the CC/CV, MSCC, TSCC/CV, CVCC/CV, CC-PC and CV-PC. The termination of the charging process is also presented. In general, the cut-off voltage for charge is used to stop the charging process if the last charging stage is in the CC period, whereas the cut-off current is used to stop the charging process if the last charging stage is in the CV period. The qualitative comparison of these charging algorithms is conducted in terms of charging time, charging efficiency, implementation complexity and cost, impact on cycle life, and sensors required.

Two recent developments in Li-ion battery charging are discussed. One is to find the optimal CCP to charge Li-ion batteries in terms of minimizing energy losses, charging time, temperature rise, or trade-off between any two of them. The other is to develop a Li-ion battery with extremely fast charging and discharging capability. The spinel LTO is used as the negative electrode to make the LTO battery instead of carbon. The LTO battery can be safely charged to 80% of the normal capacity in six minutes by using a simple CC/CC charging profile.

References

1 Chau, K.T. (2016). *Energy Systems for Electric and Hybrid Vehicles*. London: The Institution of Engineering and Technology.

2 Qiu, C., Chau, K.T., Liu, C. et al. (2014). Quantitative comparison of dynamic flux distribution of magnetic couplers for roadway electric vehicle wireless charging system. *Journal of Applied Physics* 115: 17A334-1–17A334-3.

3 Zhang, Z., Chau, K.T., Liu, C. et al. (2014). An efficient wireless power transfer system with security considerations for electric vehicle applications. *Journal of Applied Physics* 115: 17A328-1–17A328-3.

4 Berndt, D. (1997). *Maintenance-Free Batteries: Lead Acid, Nickel/Cadmium, Nickel/Metal Hydride, a Handbook of Battery Technology*, 2e. Taunton: Research Studies Press Ltd.

5 Grunditz, E.A. and Thiringer, T. (2016). Performance analysis of current BEVs-based on a comprehensive review of specifications. *IEEE Transactions on Transportation Electrification* 2 (3): 270–289.

6 Shen, W.X., Vo, T.T., and Kapoor, A. (2012). Charging algorithms of lithium-ion batteries: an overview. In: *IEEE Conference on Industrial Electronics and Applications*, 1567–1572. IEEE.

7 Jiang, J.C. and Zhang, C.P. (2015). *Fundamentals and Applications of Lithium-ion Batteries in Electric Drive Vehicles*. Singapore: John Wiley & Sons Singapore Pte. Ltd.

8 Kim, B.G., Tredeau, F.P., and Salameh, Z.M. (2008). Fast chargeability lithium polymer batteries. In: *IEEE Conference on Power and Energy Society General Meeting: Conversion and Delivery of Electrical Energy in the 21st Century*, 1–5. IEEE.

9 Chen, X.P., Shen, W.X., Vo, T.T. et al. (2012). An overview of lithium-ion batteries for electric vehicles. In: *IEEE Conference on Power and Energy*, 230–235. IEEE.

10 Panasonic Inc. (2007) Lithium-ion charging datasheet. https://na.industrial.panasonic .com/sites/default/pidsa/files/downloads/files/panasonic_li-ion_charging_information .pdf

11 Tsang, K.M. and Chan, W.L. (2011). Current sensorless quick charger for lithium-ion batteries. *Energy Conversion and Management* 52: 1593–1595.

12 Hsieh, G.C., Chen, L.R., and Huang, K.S. (2001). Fuzzy-controlled lithium-ion battery charge system with active state of charge controller. *IEEE Transactions on Industrial Electronics* 48 (3): 585–593.

13 Chen, L.R., Hsu, C.M., and Liu, C.S. (2008). A design of a grey-predicted lithium-ion battery charge system. *IEEE Transactions on Industrial Electronics* 48 (3): 3692–3701.

14 Chen, L.R. (2004). PLL-based battery charge circuit topology. *IEEE Transactions on Industrial Electronics* 51 (6): 1344–1346.

15 Chen, L.R., Chen, J.J., Chu, N.Y., and Han, G.Y. (2008). Current pumped battery charger. *IEEE Transactions on Industrial Electronics* 55 (6): 2482–2488.

16 Chan, C.C. and Chu, K.C. (1990). A microprocessor-based intelligent battery charger for electric vehicle lead acid batteries. In: *The 10th International Electric Vehicle Symposium*, 456–466.

17 Ikeya, T., Sawada, N., Takagi, S. et al. (1998). Multi-step constant-current charging method for electric vehicle, valve-regulated lead acid batteries during night time for load-levelling. *Journal of Power Sources* 75: 101–107.

18 Ikeya, T., Sawada, N., Murakami, J. et al. (2002). Multi-step constant-current charging method for an electric vehicle nickel/metal hydride battery with high-energy efficiency and long cycle life. *Journal of Power Sources* 105: 6–12.

19 Huang, J.W., Liu, Y.H., Wang, S.C., and Yang, Z.Z. (2009). Fuzzy-control-based five-step lithium-ion battery charger. In: *IEEE Conference on Power Electronics and Drive Systems*, 1547–1551. IEEE.

20 Liu, Y.H., Hsieh, C.H., and Luo, Y.F. (2011). Search for an optimal rapid charging pattern for Li-ion batteries using consecutive orthogonal arrays. *IEEE Transactions on Industrial Electronics* 26 (2): 654–661.

21 Liu, Y.H. and Luo, Y.F. (2010). Search for an optimal rapid charging pattern for Li-ion batteries using Taguchi approach. *IEEE Transactions on Industrial Electronics* 57 (12): 3963–3971.

22 Liu, Y.H., Teng, J.H., and Lin, Y.C. (2005). Search for an optimal rapid charging pattern for Li-ion batteries using ant colony system algorithm. *IEEE Transactions on Industrial Electronics* 52 (5): 1328–1336.

23 Dung, L.R. and Yen, J.H. (2010). ILP-based algorithm for lithium-ion battery charging profile. In: *IEEE International Symposium on Industrial Electronics*, 2286–2291. IEEE.

24 Vo, T.T., Shen, W.X., and Kapoor, A. (2014). New charging strategy for lithium-ion batteries based on the integration of Taguchi method and state of charge estimation. *Journal of Power Source* 273: 413–422.

25 Anseán, D., González, M., Viera, J.C. et al. (2013). Fast charging technique for high power lithium iron phosphate batteries: a cycle life analysis. *Journal of Power Sources* 239: 9–15.

26 Notten, P.H.L., Op het Veld, J.H.G., and Van Beek, J.R.G. (2005). Boostcharging Li-ion batteries: a challenging new charging concept. *Journal of Power Source* 145 (1): 89–94.

27 De Jongh, P.E. and Notten, P.H.L. (2002). Effect of current pulses on lithium intercalation batteries. *Solid State Ionics* 148: 259–268.

28 Li, J., Murphy, E., Winnick, J., and Kohl, P.A. (2001). The effects of pulse charging on cycling characteristics of commercial lithium-ion batteries. *Journal of Power Sources* 102: 302–309.

29 Chen, L.R. (2007). A design of an optimal battery pulse charge system by frequency-varied technique. *IEEE Transactions on Industrial Electronics* 54 (1): 398–405.

30 Chen, L.R. (2009). A design of duty-varied voltage pulse charger for improving lithium-ion battery-charging response. *IEEE Transactions on Industrial Electronics* 56 (2): 480–487.

31 Purushothama, B.K., Morrison, P.W., and Landau, U. (2005). Reducing mass-transport limitations by application of special pulsed current modes. *Journal of the Electrochemical Society* 152 (4): J33–J39.

32 Purushothama, B.K. and Landau, U. (2006). Rapid charging of lithium-ion batteries using pulsed current. *Journal of the Electrochemical Society* 153 (3): A533–A542.

33 Hussein, A. and Batarseh, I. (2011). A review of charging algorithms for nickel and lithium battery chargers. *IEEE Transactions on Vehicular Technology* 60: 830–838.

34 Young, K., Wang, C.S., Wang, L.Y., and Strunz, K. (2013). Electric vehicle battery technologies. In: *Electric Vehicle Integration into Modern Power Networks* (ed. R. Garcia-Valle), 15–56. New York: Springer-Verlag.

35 Klein, R., Chaturvedi, N.A., Christensen, J. et al. (2011). Optimal charging strategies in lithium-ion battery. In: *IEEE American Control Conference*, 382–387. IEEE.

36 Hu, X.S., Li, S.B., Peng, H., and Sun, F.C. (2013). Charging time and loss optimisation for LiNMC and LiFePO$_4$ batteries based on equivalent circuit models. *Journal of Power Sources* 239: 449–457.

37 Kim, N., Ahn, J.H., Kim, D.H., and Lee, B.K. (2016). Adaptive loss reduction charging strategy considering variation of internal impedance of lithium-ion polymer batteries in electric vehicles charging systems. In: *IEEE Applied Power Electronics Conference and Exposition (APEC)*, 1273–1279. IEEE.

38 Chen, Z., Xia, B., Chris Mi, C.T., and Xiong, R. (2015). Loss minimisation-based charging strategy for lithium-ion battery. *IEEE Transactions on Industry Applications* 51 (5): 4121–4129.

39 Bertsekas, D.P. and Tsitsiklis, J.N. (1995). Neuro-dynamic programming: an overview. In: *The 34th IEEE Conference on Decision and Control*, 560–564. IEEE.

40 Chen, Z., Mi, C., Xu, J. et al. (2013). Online energy management for a power-split plug-in hybrid electric vehicle based on dynamic programming and neural networks. *IEEE Transactions on Vehicular Technology* 63 (4): 1567–1580.

41 Murphey, Y.L., Park, J., Chen, Z. et al. (2012). Intelligent hybrid vehicle power control – part I: machine learning of optimal vehicle power. *IEEE Transactions on Vehicle Technology* 61 (1): 3519–3530.

42 Farmann, A., Waag, W., and Sauer, D.U. (2016). Application-specific electrical characterization of high power batteries with lithium titanate anodes for electric vehicles. *Energy* 112: 294–306.

43 Hall, F., Touzri, J., Wußler, S. et al. (2018). Experimental investigation of the thermal and cycling behavior of a lithium titanate-based lithium-ion pouch cell. *Journal of Energy Storage* 17: 109–117.

44 AA Portable Power Corp., LTO batteries. http://www.batteryspace.com/Lithium-Titanate-Battery.aspx (accessed May 2018).

45 I. Coweb, All about batteries, Part 12: Lithium titanate (LTO). 2015. https://www.eetimes.com/author.asp?section_id=36&doc_id=1325358 (accessed May 2018).

7

Battery Balancing

7.1 Background

Electric vehicles (EVs) require a battery system to provide sufficient operating voltage, power, and energy. A Li-ion battery system in EVs generally consists of many battery packs connected in series and/or parallel, and a battery pack is made of series-connected cells or parallel-connected cells. In the following, the focus is on balancing issues in the battery pack with the series-connected cells. Due to manufacturing tolerances and different operating temperatures and aging speeds, each cell in the battery pack with the series-connected cells may differ in its terminal voltage, capacity, and internal resistance and experience an unbalanced charging and discharging, where the stronger cells in the pack can be overcharged or the weaker cells in the pack can be over-discharged. After many cycles of charging and discharging, the imbalance in the battery pack deteriorates, which leads to safety issues, capacity loss, and reduced cycle life of cells in the battery pack and eventually reduced cycle life of the whole battery system. Battery balancing methods are developed to solve the unbalance problems in Li-ion battery systems. These balancing methods generally include battery sorting, passive balancing, and active balancing [1].

Battery sorting can alleviate cell imbalance in a battery pack during the initial stage of making it [2–4]. The purpose of battery sorting is to carry out screening tests for the selection of the cells which possesses similar electrochemical characteristics. The experimental data from the screening tests are obtained to cluster the cells into the group in terms of the characteristic parameters of the cell or the profiles of cell voltage or the profiles of cell temperature during the charging and discharging processes. Then, the cells in the same group are used to make a battery pack. This sorted battery pack consisting of cells with consistent characteristics can be charged or discharged as if it is a single battery cell. However, such consistency in the battery pack may not be able to be maintained throughout the lifetime due to the different operational environments, aging speeds and self-discharge rates of each cell. Battery sorting can only be useful if it is complemented by one of the other two balancing methods [5].

One of them is passive balancing. It uses a shunt resistor across each cell in the battery pack to dissipate excessive energy. Since the energy in passive balancing is converted into heat in the shunt resistor, it creates an inefficient balancing process and adds extra burden on battery thermal management. The other is active balancing. Instead of wasting the energy in the stronger cells in passive balancing, active balancing is developed

Advanced Battery Management Technologies for Electric Vehicles, First Edition. Rui Xiong and Weixiang Shen.
© 2019 John Wiley & Sons Ltd. Published 2019 by John Wiley & Sons Ltd.

to equalize the cells by transferring the excessive energy from the stronger cells to the weaker cells or to the battery pack or from the battery pack to the weaker cells [6–9]. Active balancing is able to minimize energy losses and creates an efficient balancing process.

7.2 Battery Sorting

Many methods have been developed for battery sorting. In general, they can be divided into two categories. One category is to cluster cells into the group with similar electrochemical characteristics by comparing the battery parameters (e.g. capacity and internal resistance) of each cell [4, 5]. The other category is to apply machine learning techniques (e.g. self-organizing map [SOM]) into the battery parameters of each cell for clustering the cells into the group with similar electrochemical characteristics [6–8]. In the following, a representative from each category is presented to explain battery sorting methods.

7.2.1 Battery Sorting Based on Capacity and Internal Resistance

The observation and comparison of the capacity and internal resistance of each cell is used to determine the electrochemical similarity of the cells. The measured capacity and the identified internal resistance from a battery equivalent circuit model (ECM) of each cell are studied through two screening processes in sequence [5]. These two screening process are: the capacity screening process; and the resistance screening process.

The capacity screening process is to match the capacities of the cells. The battery capacities of 20 LCO cells with nominal capacities of C_n (e.g. $C_n = 1.3$ Ah) have been tested. The constant current constant voltage (CC/CV) charging algorithm is recommended to charge the cells and the five constant currents at the discharging rates of 3C, 2C, 1C, 0.5C, and 0.2C are used to discharge the cells. Then, the average values of capacities under five discharging rates are calculated. The cells with the average capacity fall into the predefined range from C_a^u to C_a^l are selected (e.g. $C_a^l = 1.2815$ Ah and $C_a^u = 1.2864$ Ah), namely cells no. 5, no. 6, no. 13, no. 16, and no. 19, as shown in Figure 7.1. For these five cells, the open circuit voltages (OCVs) are obtained at state of charges (SOCs) of 30%, 50%, and 70% which are calculated by ampere-hour (Ah) counting. The results are listed in Table 7.1. It shows that the OCVs of the five cells are almost the same at these three specific SOCs.

For the selected five cells with similar capacities, the resistance screening process is to match DC internal resistance in the battery ECM as shown in Figure 7.2. It consists of an OCV (V_{oc}), an internal resistor R_i and a resistance–capacitor (RC) pair connected in series, which approximately reflects the dynamics of a battery.

Two resistances R_i and R_{Diff} are considered to be critical for the circuit, which are defined as DC internal resistance (R_{dc}) and can be obtained as

$$R_{dc} \approx R_i + R_{Diff} \tag{7.1}$$

The pulse discharging current (e.g. $I_{pc}^{dis} = 4$A) and the pulse charging current (e.g. $I_{pc}^{chg} = 4$A) are used to test the five cells at the specific SOCs of 30%, 50%, and 70%. Figure 7.3 shows the amplitude of current, the time duration and the voltage responses of the tests.

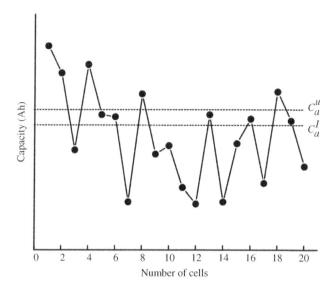

Figure 7.1 Capacities of 20 Li-ion cells for capacity screening process.

Table 7.1 OCVs (V) of five cells at SOCs of 30%, 50%, and 70%.

Battery SOCs	No. 5	No. 6	No. 13	No. 16	No. 19
30%	3.6738	3.6745	3.6751	3.6752	3.6741
50%	3.8353	3.8360	3.8358	3.8369	3.8348
70%	3.9760	3.9756	3.9773	3.9771	3.9752

Figure 7.2 Battery equivalent circuit model in charging process.

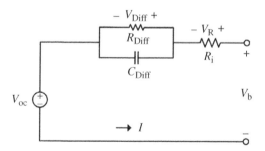

From Figure 7.3, R_i and R_{Diff} can be identified to calculate the DC internal resistances of the five cells. The results are shown in Table 7.2. It shows that the DC internal resistance of cells no. 5, no. 13, and no. 16 are almost the same. Thus, cells no. 5, no. 13, and no. 16 are grouped together to build a battery pack.

7.2.2 Battery Sorting Based on a Self-organizing Map

The SOM can be used to cluster the cells which possess similar electrochemical characteristics into a group [6]. The available capacity, internal resistance and temperature

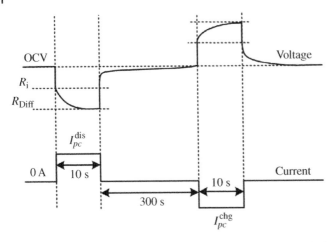

Figure 7.3 Voltage responses at a specific SOC for pulse current discharging and pulse current charging.

Table 7.2 DC internal resistances (Ω) of five cells at SOCs of 30%, 50%, and 70%.

Battery SOCs	No. 5	No. 6	No. 13	No. 16	No. 19
30%	0.07775	0.06969	0.07796	0.07757	0.08321
50%	0.07427	0.06641	0.07418	0.07462	0.08050
70%	0.07789	0.06981	0.07802	0.07812	0.08334

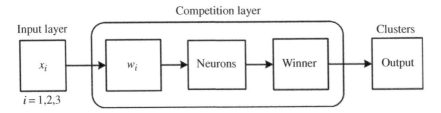

Figure 7.4 SOM for battery sorting.

variation of a cell during the discharging process are selected as the input vectors of the SOM and the output of the SOM clusters the cells into the groups. The SOM model for cell sorting is shown in Figure 7.4, where x_i ($i = 1, 2, 3$) represents the input vectors of each cell and w_i connects to each of the input vectors to the neurons which perform clustering. This SOM contains two layers: an input layer and a competition layer with a regular two-dimensional grid of mapping units.

Every unit (or neuron) i is represented by a prototype vector w_i of the same dimension as the input vectors x_i. The units are connected to the adjacent ones through a neighborhood relation. The available capacity, internal resistance and temperature variation of 12 cells are chosen as the input vectors x_i of the SOM while the outputs of the SOM are three clusters which classify 12 cells into three groups. The temperature variation has been introduced into the input vector of the SOM to achieve temperature consistency among the cells in the pack and further enhance the battery pack performance. Distances between x_i and all the prototype vectors w_i are calculated [10]. The best matching unit, which is represented by w_b, is the mapping unit with prototype closest to x_i

$$\|x_i - w_b\| = \min\{\|x_i - w_i\|\} \tag{7.2}$$

After the calculation, the prototype vectors are updated. The best matching unit and its topological neighbors are moved closer to the input vectors in the input space. The updating rule for the prototype vector of unit i is

$$w_i(t+1) = w_i(t) + \alpha(t)[x_i - w_i(t)] \tag{7.3}$$

where $\alpha(t)$ is an adaptation coefficient. The updating procedure repeats until all 12 cells are clustered successfully. The steps to implement this method are as follows:

Step 1. Conduct the experiments for 12 lithium iron phosphate (LFP) cells with each of them having the nominal capacity of 2.3 Ah under the constant discharging rate of 1C for eight times. In each time, the cells are fully charged by the CC/CV charging algorithm with a constant current of 1C and a constant voltage of 3.6 V.

Step 2. Collect the experimental data of battery available capacity, internal resistance at the fully charged state and temperature variations between the highest temperature and initial environmental temperature during the discharging process.

Step 3. Take average values of battery available capacities, internal resistances and temperature variations from eight tests for each cell, as shown in Figure 7.5.

Step 4. Regularize the averaged battery available capacity, internal resistance and temperature variation into the range of [0, 1].

Step 5. Apply the SOM to cluster the 12 cells into three groups. The results are shown in Figure 7.6.

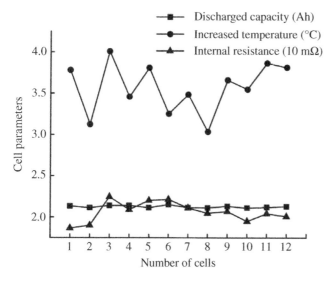

Figure 7.5 Average values of available capacity, temperature variation, and internal resistance for 12 LFP cells.

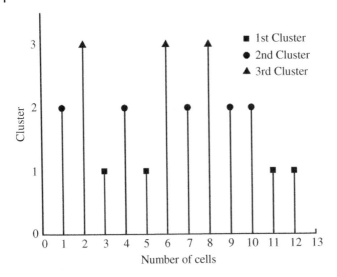

Figure 7.6 Clustering results of 12 cells.

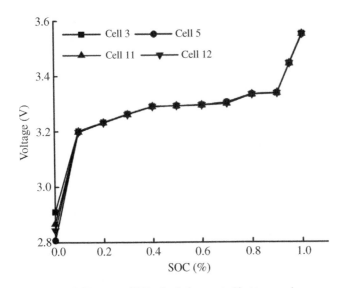

Figure 7.7 OCVs versus SOCs of cells in a sorted battery pack.

The cells in the first cluster are connected in series to build a battery pack. The experimental results of the relationship between the OCVs and the SOCs for each cell in the sorted battery pack are shown in Figure 7.7. It indicates that the cells in the same cluster have almost the same relationship of the OCVs versus the SOCs.

It can be seen from the results of two battery sorting methods that the battery sorting process is used to select cells with similar electrochemical characteristics to make the battery pack. Consequently, charging and discharging a battery pack is similar to charging and discharging a single battery cell. Furthermore, these characteristics include capacity, internal resistance and a relationship between the OCV and the SOC, which are all crucial to the accurate SOC estimation for the battery pack.

7.3   Battery Passive Balancing

Battery passive balancing releases excessive energy from stronger cells using a shunt resistor. This allows the weaker cells to catch up during the charging process so that all cells in the battery pack can be fully charged. Since no energy is saved or recovered, passive balancing is only applied in the charging process.

Generally, battery passive balancing has the following features: (i). the balancing circuit is cheap and reliable; (ii) the balancing control algorithm is simple and easy to implement; and (iii) the balancing speed is slow and directly driven by the differences among the cell terminal voltages, where the terminal voltage are normally taken as a balancing criterion. In the following, battery balancing methods based on the fixed shunt resistor, switched shunt resistor or switched transistor are presented to explain the working principle of battery passive balancing [6].

7.3.1   Fixed Shunt Resistor

The fixed shunt resistors are connected in parallel with each individual cell in a battery pack, as shown in Figure 7.8. During the charging process, the current is partially bypassed from the cells through the resistor in order to limit cell voltage. This method is continuously bypassing the current and continuously dissipates energy into heat for the cells. It is the simplest balancing method, but it is very inefficient.

7.3.2   Switched Shunt Resistor

A switch in series with a resistor is connected in parallel with each individual cell in a battery pack, as shown in Figure 7.9, where the transistor or metal–oxide–semiconductor

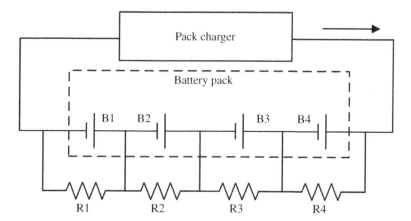

Figure 7.8 Fixed shunt resistor balancing circuit.

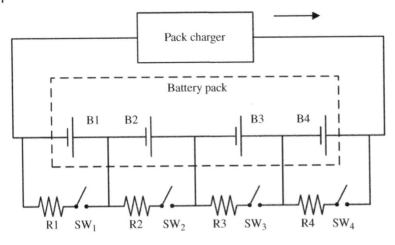

Figure 7.9 Switched shunt resistor balancing circuit.

field-effect transistor (MOSFET) can be used as the switch. There are two operation modes. The first mode turns on all switches when it is necessary. If one cell has higher voltage, more current will flow through the resistor instead of charging the cell and vice versa. With the properly selected resistance value, the weaker cells will catch up with the stronger cells effectively. The second mode involves voltage sensing and complex switch control. The voltages of each cell are sensed. The duty cycles of the switches can be regulated according to the sensed voltage levels of each cell. The higher the cell voltage, the more the energy will be dissipated. It is a simple and reliable balancing method, but it is still inefficient.

7.3.3 Shunt Transistor

A transistor is connected in parallel with each individual cell in a battery pack, as shown in Figure 7.10. During the charging process, when the cell reaches the cut-off voltage for charge, the current is proportionally bypassed around the cell to the corresponding transistor so that the cell will be charged at almost constant voltage. In this method, the

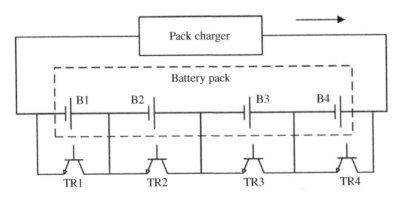

Figure 7.10 Shunt transistor balancing circuit.

current is only shunted at the end of the charging process. Compared with the switched shunt resistor working in the first mode, it has less energy loss. Compared with the switched shunt resistor working in the second mode, it does not need intelligent control, and therefore the cost is lower.

7.4 Battery Active Balancing

Battery active balancing moves energy from stronger cells or the battery pack to weaker cells in both the charging and discharging processes. This enables more energy to be delivered to the battery pack during the charging process or to be released from the battery pack during the discharging process [6–9]. Compared with passive balancing, active balancing is much more complicated and expensive but has higher balancing efficiency and faster balancing speed. The energy saving in this active balancing process may offset its complexity and high cost.

Battery active balancing involves a balancing circuit, balancing control and balancing criterion. The balancing circuit is designed to carry out the active balancing process by moving energy from stronger cells or battery pack to weaker cells. The balancing control regulates balancing current and thus balancing speed in the balancing process. The balancing criterion is used to quantify the non-uniformity of the cells in the battery pack and decides the start and termination of the balancing process.

In the following, the balancing criterion and the balancing algorithm are briefly introduced. After that, the working principles of different balancing circuits are explained and discussed in detail.

7.4.1 Balancing Criterion

Currently, battery terminal voltage, SOC or available capacity has been used as the balancing criterion in battery active balancing [11]. The battery terminal voltage is the directly measured parameter while the SOC and available capacity are the estimated parameters.

When the terminal voltage is used as a balancing criterion, the balancing process is to equalize cell terminal voltages or minimize the differences of cell terminal voltages in the battery pack to a pre-set threshold voltage. The terminal voltage as the balancing criterion is the most popular one as the voltage of each cell can be easily measured during the balancing process. However, these methods have two drawbacks: (i) cell terminal voltage does not reflect the SOC directly because it is the sum of the OCV, polarization voltage and voltage drop across internal resistance, and only the OCV has an explicit relationship with the SOC. The equalization of cell terminal voltage does not lead to a balanced SOC. (ii) LFP battery cells have flat terminal voltage in the SOC range of about 20–90%, as shown in Figure 7.11, and the battery cells in EVs normally operate in this SOC range. The voltage variation in this range is less than 0.1 V. Such slight voltage differences can hardly be used to effectively equalize the SOCs of the cells in the charging and discharging processes if the voltage measurement errors are taken into account.

To overcome the issues arising from the terminal voltage used as the balancing criterion, the SOC is selected as a balancing criterion to control the balancing process which equalizes cell SOCs or minimizes the differences of cell SOCs in the battery pack to a

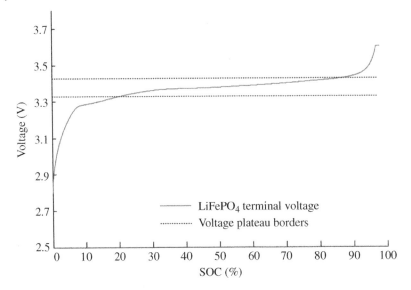

Figure 7.11 Relationship between SOC and terminal voltage for a LFP battery.

pre-set threshold SOC. However, there are still some drawbacks in real applications: (i) SOC estimation requires accurate current measurement for each cell. When a balancing circuit is not attached to the string of series-connected cells, only one current sensor is required to measure the current of each cell. When the balancing circuit is attached to the battery pack, in principle the current sensors are required to add for each cell which increases hardware complexity and implementation cost. (ii). Currently, Ah counting methods are widely used to calculate the SOCs of the cells in the balancing process, which may not be accurate in environments with high noises and uncertainties in the charging and discharging processes superimposed by the balancing process.

If the cell imbalance is caused by the available capacity difference of each cell in the battery pack due to battery aging, the above two balancing criteria are not working efficiently. The chargeable capacity (ChaC) during the charging process and the dischargeable capacity (DisC) during the discharging process are, respectively, used as a balancing criterion to equalize the cells, namely cell ChaCs are equalized or the differences of cell ChaCs are minimized to a pre-set threshold ChaC in the charging process, and cell DisCs are equalized or the differences of cell DisCs are minimized to a pre-set threshold DisC in the discharging process. The ChaC and the DisC of each cell are, respectively, defined as

$$CC_i(t) = [1 - Z_i(t)] \cdot C_i \tag{7.4}$$
$$DC_i(t) = Z_i(t) \cdot C_i \tag{7.5}$$

where $Z_i(t)$ is the dynamically estimated SOC of the ith cell and C_i is the estimated cell capacity in each charging and discharging cycle. Since the ChaC and the DisC combine the information on the SOC and the capacity of the cell, cells with different capacities in the battery pack can be equalized from the initial stage of the charging or discharging process when such differences of terminal voltages or SOCs of the cells at this stage are hardly detectable. This will provide more available balancing time during charging and discharging to maximize both the energy delivered to the cells in the charging process and the energy released from the cells in the discharging process. However, accurate estimation of battery available capacity in each charging and discharge cycle is a very challenging task.

In Section 7.5.1, a detailed discussion is provided to demonstrate that an online SOC estimation method based on a battery model can be applied to obtain the accurate SOC which is then used as a balancing criterion to equalize the battery pack.

7.4.2   Balancing Control

The balancing circuits generally provide higher balancing current at the beginning of the balancing process due to the initial large differences of terminal voltages or SOCs or available capacities of cells in the pack. When such differences decrease, balancing current reduces and so does the balancing speed. The balancing controller can regulate balancing current in an entire balancing process and thus influence balancing speed.

The existing balancing control for the balancing circuits can be categorized into two groups: open-loop balancing control; and closed-loop balancing control. In the first group, the frequency or duty cycle is predefined to control the balancing process by turning on or off the switches in the balancing circuits without any adjustment, which results in the decrease of the balancing current with the progress of the balancing process. As a result, the potential of the balancing circuits is not fully utilized [6, 11]. However, they have advantages of simplicity, low implementation complexity and high efficiency.

In the second group, the differences of the terminal voltages or SOCs or available capacities of cells in the battery pack are used as the feedback in the balancing process to adjust the frequency or duty cycle in the balancing circuits. Thus, the balancing currents can be regulated to influence the balancing speed. In Cassani and Williamson [12], the proportional–integral (PI) controller is designed to adjust the duty cycle based on the differences of the terminal voltages of the cells in a multi-switch inductor (MSI) balancing circuit, where the parameters of the PI controller are set to be constant. Due to the constant parameter settings in the PI controller, the balancing current reduces proportionally to the decrease in the voltage differences of the cells in the pack and so does the balancing speed. In Cui et al. [13], the fuzzy logic (FL) controller is designed to adjust the duty cycle in the MSI balancing circuit to increase the balancing current even when the voltage differences of the cells in the pack become smaller, leading to high level balancing current in the entire balancing process. The FL controller for the MSI balancing circuit has significantly improved the balancing speeds. The details of its design are discussed in Section 7.5.2.

7.4.3   Balancing Circuits

Battery active balancing circuits can be divided into five groups based on the direction of energy flow between cells, energy storage tank and battery pack in the balancing process, including cell to cell, cell to pack, pack to cell, cell to energy storage tank to cell, and cell to pack to cell [6–9, 11]. Theoretically, these balancing circuits can be used to equalize more than four cells in the battery pack during both charging and discharging processes, however only four cells are taken into account in the battery pack during the charging process to demonstrate their working principles for simplicity and clarity.

7.4.3.1 Cell to Cell

Many balancing circuits are developed to transfer energy from cell to cell based on the capacitor-based balancing circuits. Figure 7.12 shows the switched capacitor (SC) balancing circuit. It operates in two steps. In the first step, each capacitor is set in parallel with its corresponding upper cell, therefore the capacitor is set to the cell voltage, delivering or demanding energy from these cells. In the second step, the capacitors are set in parallel with their corresponding lower cells, transferring or demanding energy from these cells in order to reach the new voltage. After many cycles of this process, the voltage of all cells will be equalized.

The main advantages of this balancing circuit are high efficiency, low complexity and the possibility of low and high power applications. No sensing or closed-loop control is needed. The main disadvantage is the low balancing speed, as the lower the voltage difference between adjacent cells, the lower the balancing current, and therefore the lower the balancing speed. The average balancing current with the SC balancing circuit follows

$$I_{C_i} = C_i(V_{U_i} - V_{L_i})f_{sw} \tag{7.6}$$

where V_{U_i} is the higher cell voltage, V_{L_i} is the lower cell voltage and f_{sw} is the switching frequency. When the on-resistance of the switch is considered, the total resistance of the ith capacitor is

$$R_{swi} = 1/(f_{sw}C_i) + 2R_{sw}/D \tag{7.7}$$

where D is the duty cycle and R_{sw} is the on-resistance of the switches. The energy loss per step is

$$\Delta\eta = -(V_{U_i} - V_{L_i})^2/R_{sw_i} \tag{7.8}$$

To improve the balancing speed by increasing balancing current, a double-tiered switched capacitor (DTSC) balancing circuit is developed [14], as shown in Figure 7.13. For every two pairs of cells that are adjacent, one parallel capacitor (e.g. C13 or C24) is added. This design increases the balancing current and increases the balancing speed. As a result, the DTSC balancing circuit reduces the balancing time to one quarter of the SC balancing circuit. However, the addition of capacitors increases the cost.

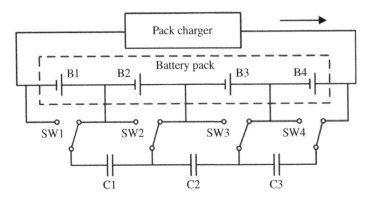

Figure 7.12 Switched capacitor balancing circuit.

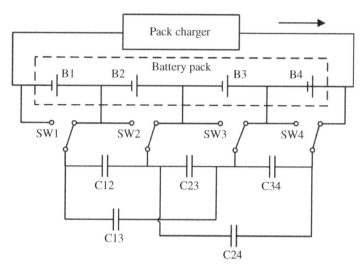

Figure 7.13 Double-tiered switched capacitor balancing circuit.

The chain structure switched capacitor (CSSC) balancing circuit is developed by adding one more capacitor to connect the top and bottom cells based on the SC balancing circuit, forming the chain structure [15], as shown in Figure 7.14. This can achieve the effective cell balancing between outer cells and reduce the steps in energy transfer when the imbalanced cells are at the ends of the string. However, this balancing circuit needs to control the direction of the energy transfer which complicates the design of the balancing controller.

The aforementioned balancing circuits with the capacitor as a main energy storage component can only use terminal voltage as a balancing criterion [11]. In these balancing circuits, the terminal voltage difference of the cells in the battery pack is the driving force that moves the energy. With the progress of the balancing process, the terminal voltage difference tends to be small and the balancing speed becomes slow. Due to the low speed,

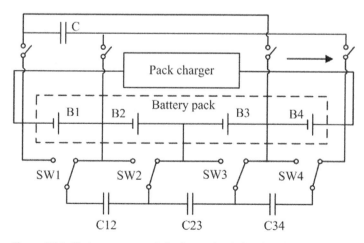

Figure 7.14 Chain structure switched capacitor balancing circuit.

these balancing circuits are generally applied to battery maintenance in stationary EVs or battery energy storage systems in power systems, where the balancing speed is not critical.

7.4.3.2 Cell to Pack

Many balancing circuits are developed to transfer energy from cell to pack. They can be implemented based on the shunt inductor, boost shunting, multiple transformers, multi-secondary windings transformer and switched transformer. Figure 7.15 shows the shunt inductor balancing circuit. It works in two steps. In the first step, when a cell is detected to have a higher voltage than the other cells of the pack, this cell will be selected in parallel with the inductor by activating the corresponding switches of the cell, i.e. the cell is shunted by the inductor, and the extra energy of the cell will be transferred into the inductor. In the second step, the switches corresponding to the higher voltage cell will be turned off while the switches SW_a and SW_b will be activated to connect the inductor to the whole battery pack, with the aim of transferring the extra energy from the higher voltage cell to the pack.

The average inductor current of a cell to be balanced is

$$I_L = [V^H_{cell_i}D^2 - V_{bat}(1-D)^2]/(2Lf_{sw}) \tag{7.9}$$

where $V^H_{cell_i}$ is the highest voltage for the cell in the battery pack, V_{bat} is the voltage of the entire battery pack, D is the duty cycle, L is the inductance, and f_{sw} is the switching frequency. The power loss can be calculated by

$$P_{loss} = 2I_L^2R_{sw} + I_L^2R_{ind} \tag{7.10}$$

where P_{loss} is the power loss, R_{sw} is the conduction resistance of the switch, and R_{ind} is the equivalent resistance of the inductor. This method is good for high power applications, but it is very slow since only one cell is being balanced at every instant.

Figure 7.16 shows the boost shunting balancing circuit [16]. It essentially consists of one master buck-boost converter with the main switch SW and one slave buck-boost converter with the switch SW_i connecting to each cell. It works in three steps. In the

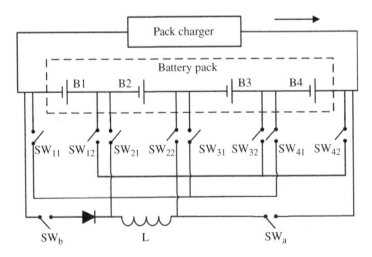

Figure 7.15 Shunt inductor balancing circuit.

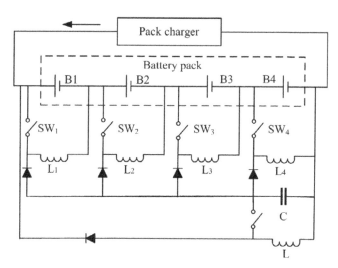

Figure 7.16 Cell to pack boost shunting balancing circuit.

first step, the switch SW_i for the highest voltage cell is turned on and the energy in this cell is transferred to the inductor L_i. In the second step, the switch SW_i is turned off and the energy in the inductor L_i is transferred to the capacitor C through the downstream cells. In the third step, the main switch SW is turned on and the energy in the capacitor C is transferred to the whole battery pack through the paralleled inductor L.

Figure 7.17 shows the multiple-transformer balancing circuit [17]. It works in two steps. In the first step, the switch SW_i of the highest voltage cell is turned on and the energy of this cell is sent to the transformer and stored in the magnetic field. In the second step, the switch SW_i is turned off and the main switch SW is turned on and the energy in the transformer is transferred to the whole battery pack. This balancing circuit is expensive since each cell needs its own transformer and switch. To reduce the cost, a multi-secondary winding transformer balancing circuit has been developed, as shown

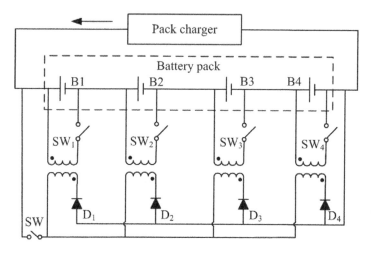

Figure 7.17 Cell to pack multiple-transformer balancing circuit.

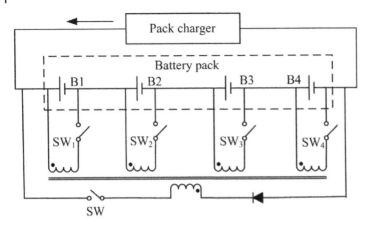

Figure 7.18 Cell to pack multi-secondary winding transformer balancing circuit.

in Figure 7.18 [18]. In this balancing circuit, each cell has its own secondary winding and all the cells share one common primary winding. It works in two steps. In the first step, the switch SW_i is turned on and the energy of the highest voltage cell is transferred and stored in the magnetic field. In the second step, the switch SW_i is turned off and the main switch SW is turned on and the energy stored in the magnetic field is fed into the whole battery pack.

Figure 7.19 shows the switched transformer balancing circuit. It connects all cells to the secondary side of a transformer through a pair of switches and free-wheeling diodes. It operates in two steps. In the first step, the energy is extracted from the highest voltage cell and stored in the transformer in the form of a magnetic field by turning on the

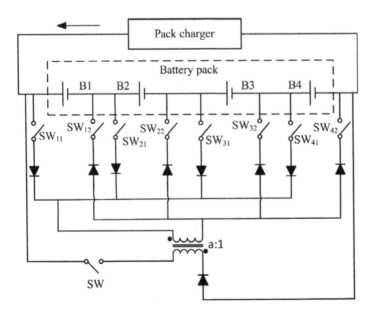

Figure 7.19 Cell to pack switched transformer balancing circuit.

corresponding two switches SW_{i1} and SW_{i2}. In the second step, the switches SW_{i1} and SW_{i2} are turned off and the main switch SW is turned on and the energy stored in the transformer is fed back into the whole battery pack.

The terminal voltage is normally taken as the balancing criterion in the above balancing circuits. However, the common feature of these balancing circuits is that only one cell with the highest voltage is selected at one time and transfers its energy into the whole battery pack. Consequently, only one current sensor is needed to measure the balancing current for each cell for the SOC estimation, where the current sensor can be serially connected to the main switch SW. Therefore, the SOC can also be used as a balancing criterion without increasing the complexity of the balancing circuits and thus the hardware cost. Since a properly designed controller can be used to regulate the balancing currents of these circuits, these balancing circuits can be applied to battery packs while EVs are moving or being charged.

7.4.3.3 Pack to Cell

Many balancing circuits are developed to transfer energy from pack to cell based on the multiple transformer, multi-secondary windings transformer and switched transformer. Figure 7.20 shows the multiple transformer balancing circuit. In this balancing circuit, the balancing operation is carried out in two steps. In the first step, the main switch SW is turned on and the energy is transferred from the battery pack to the transformer in the form of magnetic flux. In the second step, the main switch SW is turned off and the energy is sent to the cells. The cell with the lowest terminal voltage accepts the highest energy. The control of this balancing circuit is simple, but the cost is very high because each cell has a separate transformer. In this balancing circuit, the average balancing current for the selected cell is

$$I_{\text{cell}_i} = \frac{1}{2} \frac{V^L_{\text{cell}_i} D^2}{L_i f_{sw}}$$

(7.11)

where $V^L_{\text{cell}_i}$ is the lowest voltage for the cell in the battery pack, D is the duty cycle, f_{sw} is the switching frequency, and L_i is the inductor of the selected cell. To reduce the cost, the multi-secondary windings transformer balancing circuit has been developed,

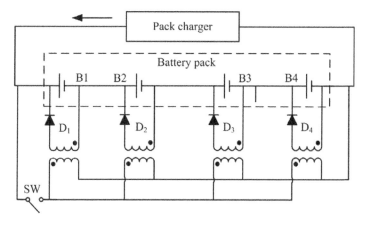

Figure 7.20 Pack to cell multiple-transformer balancing circuit.

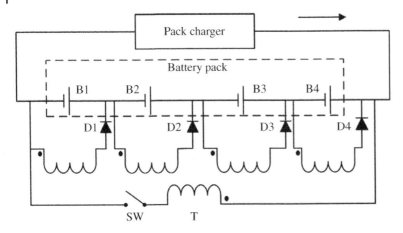

Figure 7.21 Pack to cell multi-secondary winding transformer balancing circuit.

as shown in Figure 7.21. A shared transformer has a single magnetic core with many secondary windings connecting each cell through a rectifier diode. When the imbalance is detected, the current from the battery pack flows into the transformer primary winding and induces the current in each of the secondary windings. Most of the induced current is charged to the cell with the lowest voltage.

Similar to the multi-secondary windings transformer, the switched transformer balancing circuit has been developed, as shown in Figure 7.22. It works in two steps. In the first step, the main switch SW on the primary side is turned on and the energy from the battery pack is transferred to the transformer in the form of magnetic flux. In the second step, the main switch SW is turned off and the two switches SW_{i1} and SW_{i2} are turned

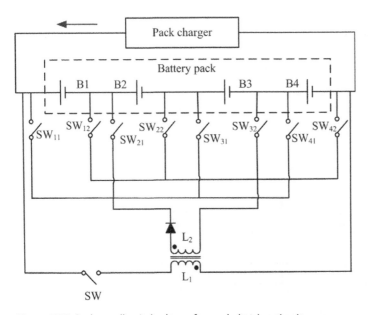

Figure 7.22 Pack to cell switched transformer balancing circuit.

on to connect the selected cell which has the lowest voltage, which allows the energy to transfer to the selected cell.

7.4.3.4   Cell to Energy Storage Tank to Cell

The balancing circuits are also developed to transfer energy from cell to energy storage tank and then to cell (abbreviated below as cell to tank to cell). They can be implemented based on the single switched inductor and single switched capacitor. Figure 7.23 shows the single switched inductor balancing circuit [19], where the inductor L is utilized as the energy storage tank. In this balancing circuit, the controller only selects the switches that connect the inductor in parallel with the strongest cell (i.e. the highest voltage cell) and transfers the energy to the inductor. Then, the controller selects the switches that connect the inductor in parallel with the weakest cell (i.e. the lowest voltage cell), transferring the excessive energy in the strongest cells through the inductor to the weakest cell. This balancing circuit can connect each cell to the central inductor through the switch matrix and transfers energy among any two cells. It has relatively low cost and high efficiency for high power applications. But, it is slow balancing as only one cell is balanced at the same time. The average balancing current and the energy loss for the selected cell can be calculated using Eqs. (7.9) and (7.10), respectively.

Similar to the single switched inductor balancing circuit, a single switched capacitor is developed to transfer energy from cell to tank to cell, as shown in Figure 7.24. Instead of using an inductor, a capacitor is utilized as the energy storage tank to store the energy and transfer the energy between the strongest cell (i.e. the highest voltage cell) and the weakest cell (i.e. the lowest voltage cell).

7.4.3.5   Cell to Pack to Cell

Balancing circuits are further developed to transfer energy from cell to pack to cell based on the bidirectional switched transformer, bidirectional multi-secondary winding transformer and bidirectional multiple transformer (BMT) balancing circuits. Figure 7.25

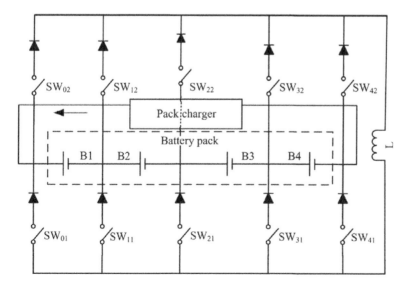

Figure 7.23 Cell to tank to cell single switched inductor balancing circuit.

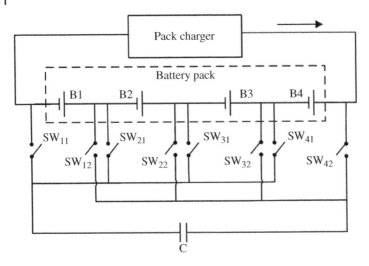

Figure 7.24 Single switched capacitor balancing circuit.

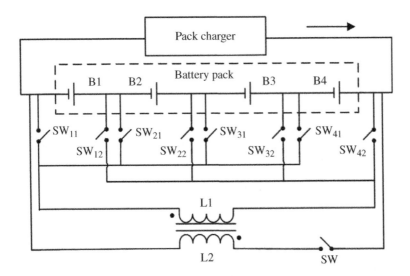

Figure 7.25 Cell to pack to cell bidirectional switched transformer balancing circuit.

shows the bidirectional switched transformer balancing circuit. This balancing circuit can work in three modes: namely cell to pack, pack to cell, or cell to pack to cell. A single transformer is utilized to transfer the energy from the strongest cell to the pack or from the pack to the weakest cell or from the strongest cell to the pack to the weakest cell by selecting the corresponding switches from the switch matrix.

Figure 7.26 shows the BMTs balancing circuit [20]. This balancing circuit allows the transfer of the energy from each cell to the whole battery pack and vice versa. If the battery pack is in the charging state, the excess energy from the strong cells can be transferred to the whole battery pack. If the battery pack is in the discharging state, the energy from the battery pack can be transferred to the weak cells. Figure 7.27

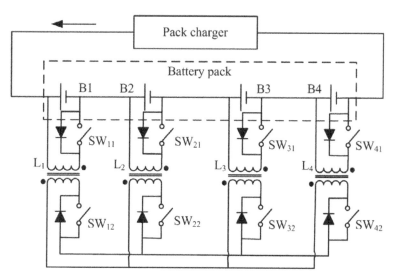

Figure 7.26 Cell to pack to cell bidirectional multiple transformers balancing circuit.

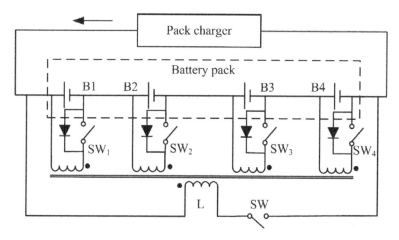

Figure 7.27 Cell to pack to cell bidirectional multi-secondary windings transformer balancing circuit.

shows the bidirectional multi-secondary windings transformer (BMSWT) balancing circuit [21].

In this balancing circuit, instead of all cells having their own primary windings in the BMTs balancing circuit, all cells share the common primary winding. The working principle of the BMSWT balancing circuit is similar to the BMTs balancing circuit.

7.5 Battery Active Balancing Systems

Two active balancing systems are presented to demonstrate how they equalize a battery pack for capacity improvement from the perspectives of selecting a balancing criterion and designing a balancing controller [22, 23].

7.5.1 Active Balancing System Based on the SOC as a Balancing Criterion

7.5.1.1 Battery Balancing Criterion

The adaptive extended Kalman filter (AEKF) is applied for the real-time SOC estimation for the battery pack, which is used as a balancing criterion [23]. A battery pack is considered to be a dynamic system, and its state space equation can be generalized as

$$\begin{cases} x_k = f(x_k, u_k, \omega_k) \\ y_k = h_k(x_{k,} v_k) \end{cases} \tag{7.12}$$

$$\begin{cases} \omega_k \sim (0, Q_k) \\ v_k \sim (0, R_k) \end{cases} \tag{7.13}$$

where x_k is a state which can be written as a vector $[Z_k \ V_{pc,k} \ V_{pe,k}]^T$, y_k is an output which represents a battery terminal voltage, ω_k is the process Gaussian noises with the covariance of Q_k, v_k is measurement Gaussian noises with the covariance of R_k, which represent process errors and measurement noises of a battery system, respectively. $V_{pc,k}$ and $V_{pe,k}$ are the polarization voltages in the battery ECM as shown Figure 7.28.

In the ECM, V_t and I represent the battery terminal voltage and current, respectively; R_{in} is an internal resistance characterizing the instant voltage drop; a parallel branch of a resistance R_{pe} and a capacitance C_{pe} represents electrochemical polarization and the other parallel branch represents concentration polarization; these two branches reflect the short-term and long-term transient responses of the battery. The symbols ΔVoc, ΔR_{in}, ΔR_{pe}, ΔC_{pe}, ΔR_{pc}, and ΔC_{pc} are associated with the errors and noises of a battery system. The capacitance C_n represents the total charge stored in the battery by converting the capacity in ampere-hour (Ah) into the charge in Coulomb

$$C_n = 3600 \cdot C_{bat} \tag{7.14}$$

where C_{bat} is the battery capacity in Ah. The voltage V_{SOC} across the capacitance C_n represents the SOC and its value is set between 0 and 1 V corresponding to 0% and

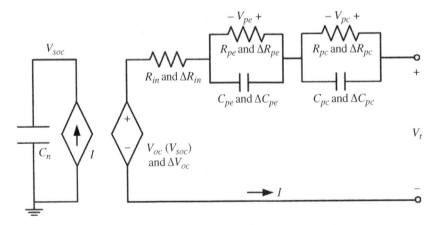

Figure 7.28 Schematic diagram of battery equivalent circuit model in charging process.

100%. The SOC is defined as

$$Z(t) = Z(0) + (1/C_n) \int_{t_0}^{t} I(\tau)d\tau \tag{7.15}$$

Based on the ECM, the state space equation for this battery pack in terms of the battery parameters is derived, and a detailed discussion can be found in Chapter 2.

The Jacobian matrixes of partial derivatives from the state space equation are as follows:

$$A_{k-1} = \left.\frac{\partial f}{\partial x}\right|_{\hat{x}_{k-1}}, B_{k-1} = \left.\frac{\partial f}{\partial u}\right|_{\hat{x}_{k-1}}, H_k = \left.\frac{\partial h}{\partial x}\right|_{\hat{x}_k} \tag{7.16}$$

Then the linearized state equation including noises could be generalized as

$$\begin{cases} x_k = A_{k-1}x_{k-1} + B_{k-1}u_{k-1} + \omega_{k-1} \\ y_k = H_k x_k + v_k \\ E(\omega_k \omega_j^T) = Q_k \delta_{k-j} \\ E(v_k v_j^T) = R_k \delta_{k-j} \\ E(\omega_k v_j^T) = 0 \end{cases} \tag{7.17}$$

The steps for the SOC estimation with the AEKF can be summarized as follows:

Step 1: *Initialization.* The initial x_0 is estimated as a Gaussian random vector with mean of x_0 and covariance of P_0.

$$\begin{cases} \hat{x}_0^+ = E[x_0] \\ P_0^+ = E[(x_0 - \hat{x}_0^+)(x_0 - \hat{x}_0^+)^T] \end{cases} \tag{7.18}$$

Step 2: *Time update* (from time $(k-1)^+$ to time k^-). The current state estimation is obtained based on the state estimation and its covariance in the previous step.

$$\begin{aligned} \widetilde{P}_k^- &= \alpha^2 A_{k-1}\widetilde{P}_{k-1}^+ A_{k-1}^T + Q_{k-1} \\ \hat{x}_k^- &= A_{k-1}\hat{x}_{k-1}^+ + G_{k-1}u_{k-1} \end{aligned} \tag{7.19}$$

where $\widetilde{P}_k^- = \alpha^{2k}P_k^-$, $\widetilde{P}_{k-1}^+ = \alpha^{2(k-1)}P_{k-1}^+$, and α is the fading memory factor and its value is set as 1.0001 in this study. When the fading memory factor is larger than 1, it serves to provide less credence to the SOC estimation and more credence to the measurement [24].

Step 3: *Kalman gain update.*

$$K_k = \widetilde{P}_k^- H_k^T (H_k \widetilde{P}_k^- H_k^T + R_k)^{-1} \tag{7.20}$$

Step 4: *Discrete measurement update.* The estimated state $\hat{x}_k^-$ is updated when the measurement is available. The covariance is also updated and they are presented by

$$\begin{aligned} \hat{x}_k^+ &= \hat{x}_k^- + K_k(y_k - h_k(\hat{x}_k^-, 0, t_k)) \\ \widetilde{P}_k^+ &= (I - K_k H_k)\widetilde{P}_k^-(I - K_k H_k)^T + K_k R_k K_k^T \end{aligned} \tag{7.21}$$

where $\widetilde{P}_k^+ = \alpha^{2k}P_k^+$.

The initial parameters of the AEKF are obtained based on empirical experience. For example, the initial parameters of the battery cell for the pulse constant current (PCC) test are tuned to be

$$x_0 = [0.1\ 0.01\ 0.01], R_0 = [20], Q_0 = \begin{bmatrix} 0.05 & 0 & 0 \\ 0 & 0.2 & 0 \\ 0 & 0 & 0.2 \end{bmatrix}, P_0 = \begin{bmatrix} 100 & 0 & 0 \\ 0 & 0.01 & 0 \\ 0 & 0 & 0.01 \end{bmatrix}.$$

The model parameters in the ECM as shown in Figure 7.28 are obtained by fitting the experimental data from the PCC test as shown in Figure 7.29. A LFP battery (A123 ANR26650) with nominal capacity of 2.3 Ah is tested under room temperature. The cell is first fully discharged with 2.3 A until the voltage reaches 2 V. Then, the cell is rested for one hour to get the initial OCV. After that, the current profile of the PCC as shown in Figure 7.29a is used to charge the battery, where one-hour rest is inserted to obtain the OCV for every 10% SOC increment in the charging process. Figure 7.29b shows the transient voltage corresponding to the PCC. From Figure 7.29a,b, the relationship between the OCV and the SOC is obtained and shown in Figure 7.30. This OCV data can be fitted by

$$V_{oc}(Z) = \lambda_0 + \lambda_1 \cdot Z + \lambda_2 \cdot Z^2 + \lambda_3 \cdot Z^3 + \lambda_4 \cdot Z^4 + \lambda_5 \cdot Z^5 + \lambda_6 \cdot Z^6 + \lambda_7 \cdot Z^7 \quad (7.22)$$

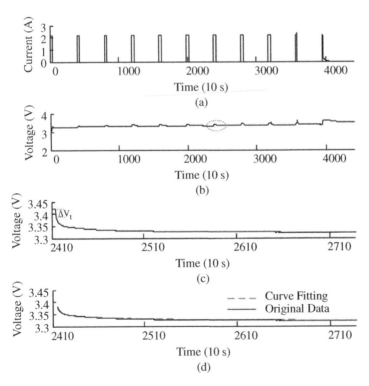

Figure 7.29 Experimental results of the PCC test: (a) pulse charge currents; (b) terminal voltage responses; (c) zoomed-in transient terminal voltage response corresponding to seventh current pulse; (d) comparison of curve fitting and experimental results.

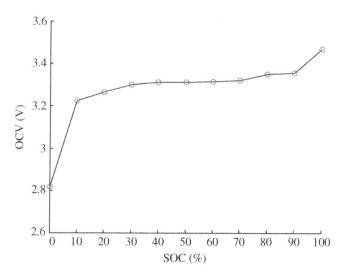

Figure 7.30 OCV–SOC relationship for a LFP battery from the PCC test.

Table 7.3 Coefficients for relationship between SOC and OCV of a battery cell.

Parameters	λ_0	λ_1	λ_2	λ_3	λ_4	λ_5	λ_6	λ_7
Values	2.82	9.43	−84.32	380.60	−927.90	1240.00	−854.20	237.10

where $\lambda_i(i = 0, \ldots, 7)$ are the coefficients of the OCV versus the SOC [22] and the values are listed in Table 7.3.

The part circled in Figure 7.29b is shown zoomed-in in Figure 7.29c. Since the battery model parameters corresponding to the pulse at the SOC of 70% cause the highest SOC estimation error, the parameters obtained at this SOC will be applied in the model for the SOC estimation. In Figure 7.29c, the relaxation voltage can be intuitively represented by

$$V_t(t) = V_{oc} - V_{pe}\exp(-t/\tau_{pe}) - V_{pc}\exp(-t/\tau_{pc}) \tag{7.23}$$

ΔV_t represents the voltage drop caused by the internal resistance R_{in}, so it can be calculated by

$$R_{in} = \Delta V_t/I \tag{7.24}$$

Then, a root mean square error method is used to determine V_{pe}, V_{pc}, τ_{pe}, and τ_{pc} in Eq. (7.21), the resistances and capacitances in the two parallel branches are calculated by

$$R_{pe} = V_{pe}/I, R_{pc} = V_{pc}/I, C_{pe} = \tau_{pe}/R_{pe}, C_{pc} = \tau_{pc}/R_{pc} \tag{7.25}$$

All these parameters are shown in Table 7.4. They are substituted into Eq. (7.21) to calculate the transient voltage, which is then compared with the experimental results in Figure 7.29d. It shows a good agreement.

Table 7.4 Parameters for equivalent circuit model of a battery cell.

Parameters	C_n(F)	R_{in}(mΩ)	C_{pe}(F)	R_{pe}(mΩ)	C_{pc}(F)	R_{pc}(mΩ)
Values	7760	16	3958	15.3	86 094	9

7.5.1.2 Battery Balancing Circuit

A flyback converter balancing circuit is chosen and the method to obtain cell current is designed as shown in Figure 7.31 [25]. In this balancing circuit, only one additional current sensor is added to calculate the current ($I_i(t)$) of each cell in the battery pack during the balancing operation, namely the balancing current sensor measures the balancing current (I_{b1}), the existing charging current sensor measures the charging current ($I(t)$), then the current of each cell ($I_i(t)$) is calculated by

$$I_i(t) = \begin{cases} I(t) - I_{b1} + I_{b2} & \text{Balanced cell} \\ I(t) + I_{b2} & \text{Other cells} \end{cases} \tag{7.26}$$

The balancing current sensor measures the balancing current in the low voltage side to improve the accuracy of current measurement and the current (I_{b2}) in the high voltage side is calculated by

$$I_{b2} = (N_1/N_2) \cdot I_{b1} \cdot \eta \tag{7.27}$$

where N_1 and N_2 are the ratio of the flyback transformer and η is the efficiency of the flyback converter. In the balancing operation, the switches are turned on/off at high frequency, the average balancing current of each cell is superimposed to the charging current to obtain $I_i(t)$, which is used in the AEKF for the SOC estimation in real-time.

This balancing circuit can select any cells in the pack for the balancing operation through a switch matrix in two operation modes: the cell to pack mode (CTP) and the pack to cell mode (PTC). In the CTP mode, the cell with the highest SOC or terminal voltage is discharged and releases excessive energy back to the pack via the flyback converter. In the PTC mode, the cell with the lowest SOC or terminal voltage is identified and the energy from the pack is injected into this cell.

7.5.1.3 Battery Balancing Control

The open loop control is applied in this balancing circuit. The duty cycle and frequency of the PWM control signal for the MOFETs are set to be constant, which are 0.4 and 10 kHz, respectively. The sampling frequency of the current sensor is set to 50 kHz. The converter is running in discontinuous conduction mode with the peak balancing current limited to 1.5 A. The inductor values of the transformer are 70 μH and 930 μH for the low and high voltage sides, respectively.

7.5.1.4 Experimental Results

The active balancing system based on the SOC as a balancing criterion is evaluated and compared with the active balancing system based on the battery terminal voltage as a balancing criterion [22]. Three A123 LFP battery cells are connected in series to make

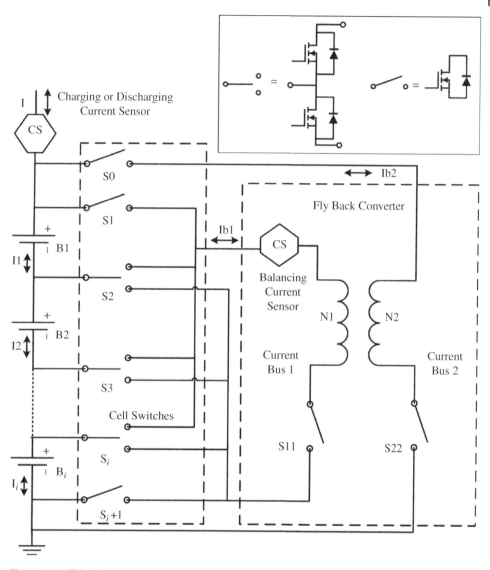

Figure 7.31 Flyback converter balancing circuit with battery pack.

a battery pack. Their specifications are listed in Table 7.5. As a trade-off between charged pack capacity and charging time, the battery pack stops charging when the voltage of any battery cell reaches the upper cut-off voltage (e.g. 3.6 V for the selected battery cell). The initial SOCs of cell one, two, and three are, respectively, set to 0%, 0%, and 20% to create an imbalance scenario. Figure 7.32 shows the experimental results for the active balancing system based on the SOC as a balancing criterion. The estimated SOC values of these three cells are shown in Figure 7.32a. It can be observed that the SOC difference continues converging with increasing balancing operation time, dropping from the highest SOC difference at the beginning to less than 2% after 2700 seconds.

Table 7.5 Specification of three LFP battery cells in the pack.

Battery type	Cell one	Cell two	Cell three
Nominal capacity (Ah)	2.3	2.3	2.3
Tested capacity (Ah)	2.11	2.16	2.17
Internal resistance (mΩ)	20	16	20

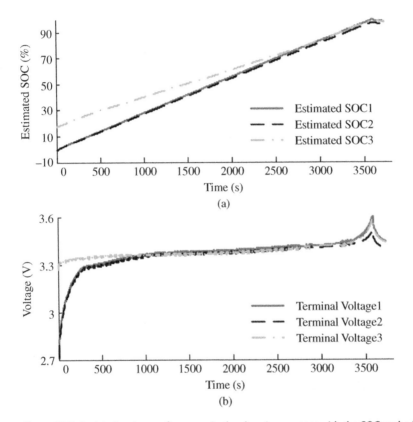

Figure 7.32 Pack balancing performance in the charging process with the SOC as the balancing criterion. (a) Experimental results for estimated SOCs of three cells during the charging process. (b) Experimental results for terminal voltages of three cells during the charging process.

The terminal voltages of these three cells are shown Figure 7.32b. Since cell three has the highest initial SOC, its extra charge is discharged to the pack in the balancing process, and it therefore has the lowest terminal voltage in the middle part (the plateau) of the charging process.

Figure 7.33 shows the experimental results for the active balancing system based on the terminal voltage as a balancing criterion. Considering the accuracy of the voltage sensor and the hysteresis of battery cells, the balancing operation starts when the maximum terminal voltage difference is higher than 20 mV is adopted to avoid

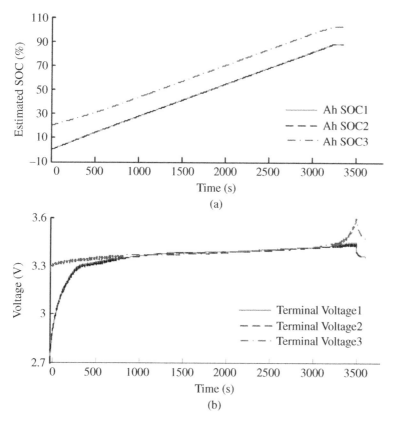

Figure 7.33 Pack balancing performance in the charging process with terminal voltage as the balancing criterion. (a) Experimental results for Coulomb counting SOCs of three cells during the charging process. (b) Experimental results for terminal voltages of three cells during the charging process.

unnecessary balancing operation. Figure 7.33a shows the SOC values calculated using the Coulomb counting method during the charging process. The SOC difference between cell three and the other two cells remains large during the entire charging process. The terminal voltages of the three cells are shown in Figure 7.33b, indicating that there is no balancing operation during most of the charging process. The balancing happens at the beginning when the terminal voltage of cell three is the highest and the maximum voltage difference is larger than 20 mV. Then, the terminal voltages of cell one and cell two rise quickly in the initial charging stage and reach the terminal voltage of cell three at around 700 seconds and the balancing stops since the maximum voltage difference in the pack becomes less than 20 mV. The balancing starts again near the end of the charging process. During the voltage plateau, the terminal voltage of cell three is slightly lower than the other two cells, even though its SOC is still the highest.

The balancing results with these two balancing criteria are compared in Table 7.6. When the SOC is used as a balancing criterion, the SOC values of these three cells are 97.9%, 96.7%, and 97.6% at the end of charging, respectively. This shows that the battery pack is well balanced and the charged pack capacity is 2.07 Ah. When the terminal

Table 7.6 Comparison of active balancing systems using different balancing criteria.

Balancing criteria	SOCs (%)			Charged pack capacity (Ah)
	Cell one	Cell two	Cell three	
SOC	97.9	96.7	97.6	2.07
Terminal voltage	86.8	85.1	98.4	1.83

voltage is used as a balancing criterion, the SOC values are 86.8%, 85.1%, and 98.4% at the end of charging, respectively. This shows that the battery pack is still unbalanced in terms of the SOC. This is due to the fact that the terminal voltages of the three cells are very close to each other in the plateau, the small voltage difference cannot trigger the balancing process for most of the charging time. Since the cell with the lowest SOC decides the pack capacity, it is found that the charged pack capacity is only 1.83 Ah, which is much less than 2.07 Ah.

7.5.2 Active Balancing System Based on FL Controller

A FL controller is designed to control a MSI balancing circuit for the battery pack equalization, as shown in Figure 7.34 [23]. The MSI balancing circuit operates at different modes [12]. In each mode, one cell is disconnected from the balancing circuit by turning off the MOSFET. Such off-state duration, which is also called the mode duration, affects the direction and amplitude of balancing currents.

In the following, a four-cell battery pack is taken as an example to explain the MSI balancing circuit, which comprises three inductors and four power MOSFETs. There are four control signals for the MOSFETs, as shown in Figure 7.35, resulting in four working modes, as shown in Figure 7.36. A short DT) is inserted in each mode to avoid short circuits.

Assume that the MSI balancing circuit operates in the steady state, according to Kirchhoff's current law, the relationship between the cell balancing current and the mode duration can be expressed as

$$
\begin{pmatrix} I_{b1} \\ I_{b2} \\ I_{b3} \\ I_{b4} \end{pmatrix} = \begin{bmatrix} -(d_2 + d_3 + d_4) & -(d_3 + d_4) & -d_4 \\ d_1 & -(d_3 + d_4) & -d_4 \\ d_1 & d_1 + d_2 & -d_4 \\ d_1 & d_1 + d_2 & d_1 + d_2 + d_3 \end{bmatrix} \cdot \begin{pmatrix} I_{L1} \\ I_{L2} \\ I_{L3} \end{pmatrix} \cdot \left(\frac{1}{T} \right) \qquad (7.28)
$$

where d_x represents the mode duration in percentage of one period for mode x ($x = 1,2,3,4$) and the sum of d_x should be equal to one. T is the switching period. It can be seen from Eq. (7.28) that the cell balancing current is decided by both the inductor current and the mode duration.

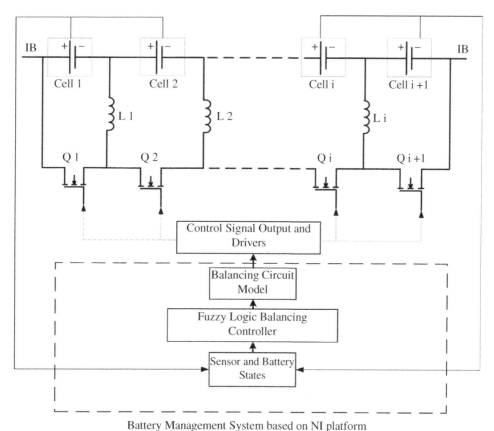

Figure 7.34 MSI balancing system with FL controller.

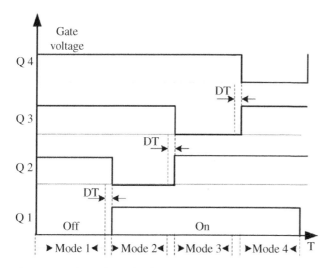

Figure 7.35 Control signals of MOSFET. DT, dead-time.

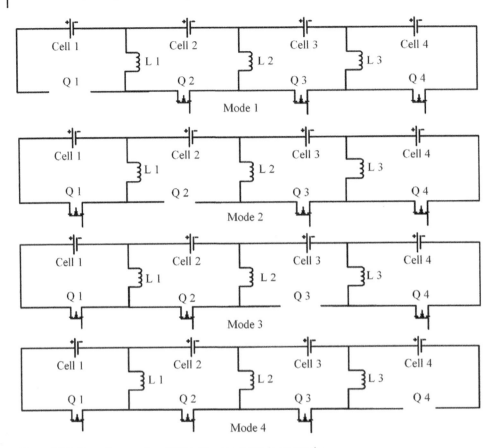

Figure 7.36 Operation modes of MSI balancing circuit in one cycle.

According to the volt-second balance, the relationship between the cell voltage and the mode duration can be expressed as [26]:

$$\begin{cases} (V_{b2} + V_{b3} + V_{b4}) \cdot d_1 \cdot T + (-V_{b1}) \cdot (d_2 + d_3 + d_4) \cdot T = 0 \\ (V_{b3} + V_{b4}) \cdot (d_1 + d_2) \cdot T + (-V_{b1} - V_{b2}) \cdot (d_3 + d_4) \cdot T = 0 \\ V_{b4} \cdot (d_1 + d_2 + d_3) \cdot T + (-V_{b1} - V_{b2} - V_{b3}) \cdot d_4 \cdot T = 0 \end{cases} \quad (7.29)$$

In relation to the inductor resistance (R_L) and the MOSFET conduction resistance (R_{on}) in the MSI balancing circuit, the relationship between the inductor balancing current and the MOSFET mode duration is given by [24]:

$$\begin{pmatrix} \dfrac{V_{b1}}{R_{on}} + I_{L1} + I_{L2} + I_{L3} + I_{L1}\dfrac{R_L}{R_{on}} \\[2mm] \dfrac{V_{b1} + V_{b2}}{R_{on}} + I_{L1} + 2I_{L2} + 2I_{L3} + I_{L2}\dfrac{R_L}{R_{on}} \\[2mm] \dfrac{V_{b1} + V_{b2} + V_{b3}}{R_{on}} + I_{L1} + 2I_{L2} + 3I_{L3} + I_{L3}\dfrac{R_L}{R_{on}} \end{pmatrix}$$

$$= \begin{pmatrix} \dfrac{V_p}{R_{on}} - 2I_{L1} - I_{L2} & I_{L2} + I_{L3} & I_{L3} \\[2mm] \dfrac{V_p}{R_{on}} - I_{L1} + I_{L3} & \dfrac{V_p}{R_{on}} + I_{L1} + I_{L3} & 2I_{L3} \\[2mm] \dfrac{V_p}{R_{on}} + I_{L2} + 2I_{L3} & \dfrac{V_p}{R_{on}} + I_{L1} + I_{L2} + 2I_{L3} & \dfrac{V_p}{R_{on}} + I_{L1} + 2I_{L2} + 2I_{L3} \end{pmatrix}$$

$$\cdot \begin{pmatrix} d_1 \\ d_2 \\ d_3 \end{pmatrix} \tag{7.30}$$

where V_p is the pack voltage.

7.5.2.1 Balancing Principle

The MSI balancing circuit uses the inductor to transfer the energy in the cell by switching MOSFETs in sequence [23]. The direction of the energy transfer is decided by both the cell voltage and the mode duration. By changing the ratio of mode duration to cell voltage (RMV), the cell balancing current direction can be controlled. If the RMVs of all the cells are the same, the average cell balancing current is zero. Consider cell one in the pack as an example. When the RMV of cell one is lower than that of the others, this cell is charged. When the RMV of cell one is higher than that of the others in the pack, this cell is discharged. Mathematically, it can be expressed as

$$\begin{cases} \dfrac{d_1}{V_{b1}} = \dfrac{d_2}{V_{b2}} = \dfrac{d_3}{V_{b3}} = \dfrac{d_4}{V_{b4}} & \text{No cell is balanced} \\[3mm] \dfrac{d_1}{V_{b1}} > \dfrac{d_2}{V_{b2}} = \dfrac{d_3}{V_{b3}} = \dfrac{d_4}{V_{b4}} & \text{Cell one is charged} \\[3mm] \dfrac{d_1}{V_{b1}} < \dfrac{d_2}{V_{b2}} = \dfrac{d_3}{V_{b3}} = \dfrac{d_4}{V_{b4}} & \text{Cell one is discharged} \end{cases} \tag{7.31}$$

Based on this principle, the MSI balancing circuit can be controlled by constant equal mode duration and the simulation results are shown in Figure 7.37. It can be seen that the balancing current rises to the peak which may be too high for the battery and then drops quickly when the voltage difference decreases. To improve balancing performance, the FL controller is designed to control the MSI balancing circuit by changing mode duration, leading to the possibility of regulating the amplitude and direction of the balancing current over the whole balancing process.

7.5.2.2 Design of FL Controller

A FL controller is the application of the fuzzy theory [27]. It consists of three parts: fuzzification, fuzzy inference, and defuzzification (Figure 7.38). In fuzzification, crisp inputs are fuzzified into linguistic variables using membership functions (MFs). In fuzzy inference, there are two parts: an inference engine; and a rule base. The inference engine decides the FL operations and the rule base stores the control rules based on expert knowledge. The linguistic control outputs are generated by the inference engine. In defuzzification, the linguistic outputs are converted back to the crisp output using the

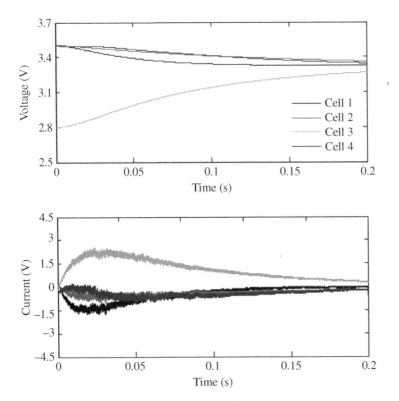

Figure 7.37 Simulation results for MSI balancing circuit with constant equal mode duration.

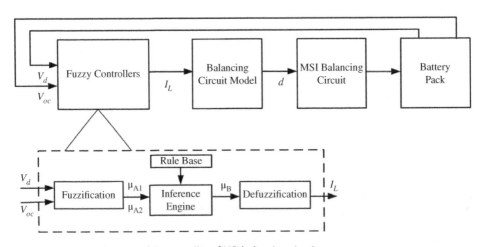

Figure 7.38 Block diagram of FL controller of MSI balancing circuit.

center of gravity method. The FL controller is designed to control the inductor current (I_L) through changing the mode duration (d) which is the time to turn the MOSFETs on/off in the MSI balancing circuit.

The key to the FL controller is the design of its MFs [28]. A MF defines how each point of the variables in the input space is mapped to a degree of membership between 0 and 1. There are two fuzzy input variables, cell OCV (V_{oc}) and cell OCV difference (V_d), and one fuzzy output variable, inductor current (I_L). The OCV difference of cell i is defined as:

$$V_{di} = V_{oc_i} - V_{oc_{i+1}} \tag{7.32}$$

A LFP battery cell is used as an example to explain the design of MF. The OCV of the LFP battery cell is described by five linguistic variables, i.e. VS (very small), S (small), M (medium), L (large), and VL (very large) in triangular and trapezoidal forms, as shown in Figure 7.39. Since the LFP battery has a very flat voltage around 3.3 V, the MFs concentrate on 3.3 V. The OCV difference V_d and the inductor current I_L are described by the same five linguistic variables as the OCV, i.e. NV (negative large), N (negative), Z (around zero), P (positive), and PV (positive large), as shown in Figures 7.40 and 7.41, respectively. The V_d range covers all the possible cell voltage differences in the pack. However, in most cases these difference values are between −0.1 V and 0.1 V, which is shown in the enlarged figure in Figure 7.40. Since the sum of the MFs at one point is not required to be 1 in the FL system [29], the Gaussian MF is applied to the linguistic variable Z for the MFs of V_d and I_L. This causes the output inductor current to drop quickly

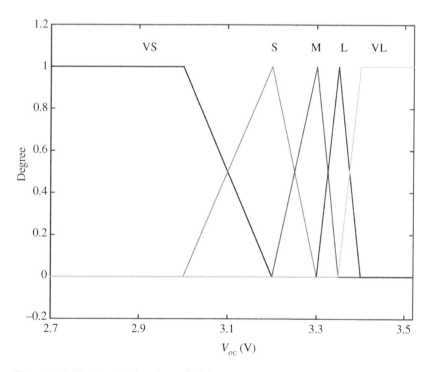

Figure 7.39 Membership functions of OCVs.

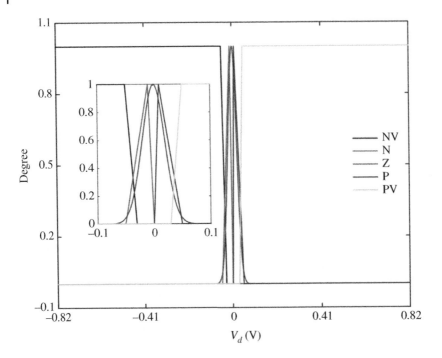

Figure 7.40 Membership function of OCV differences.

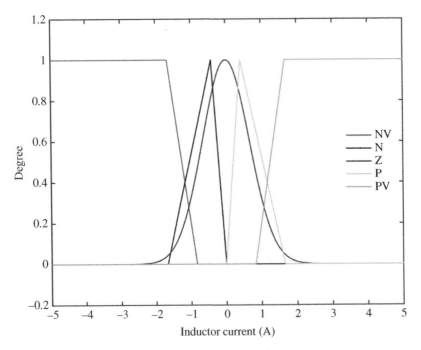

Figure 7.41 Membership function of inductor currents.

to near zero when V_d is less than 20 mV, reducing the balancing loss and the chance of divergence.

In the following, the design of the FL controller is explained and the procedure is summarized as follows [29]:

Step 1: Determination of fuzzy rules. The rules describe the knowledge about the behavior of a complex system. These rules with two inputs can be generally expressed as:

$$R^k : \text{If } X_1 \text{ is } A_1^k \text{ and } X_2 \text{ is } A_2^k, \text{ then } Y \text{ is } B^k \quad \text{for } k = 1, 2, \dots, n. \tag{7.33}$$

where n is the total number of the fuzzy rules and R^k denotes the kth rule. A_1^k, A_2^k ($k = 1, 2, \dots, n$), and B^k ($k = 1, 2, \dots, n$) are the fuzzy sub-sets of X_1, X_2, and Y, respectively. In this study, X_1 and X_2 are input linguistic variables V_d and V_{oc}, respectively, and Y is output linguistic variable I_L. The detailed fuzzy rules are given in Table 7.7. These rules describe the relation between input and output of the FL controller based on the expert knowledge of LFP battery balancing. There is a total of 25 rules. The following are two of the rules.

Rule 01: IF $V_d = \text{NV}$ and $V_{oc} = \text{VS}$ THEN $I_L = \text{NV}$;
Rule 25: IF $V_d = \text{PV}$ and $V_{oc} = \text{VL}$ THEN $I_L = \text{PV}$;

Step 2: Fuzzification of the input variables. Fuzzy sets for the variables are determined by the MFs. The crisp inputs are converted into the degree of membership between 0 and 1 using the following equation:

$$x_i = \mu_{A_j^k}(x_i) \tag{7.34}$$

where x_i is the input value of ith input variable, and $\mu_{A_j^k}(x)$ is the fuzzy MF of the input linguistic variables V_d and V_{oc}.

Step 3: Fuzzy inference. This step applies fuzzy rules to map the given inputs to an output fuzzy set with the FL operations. Max-min composition is utilized. First, the output fuzzy set of each rule with implication operation is computed, and then all the output sets are combined into a single fuzzy set with aggregation operation. Mathematically, the process can be written as

$$u_{B^k}(I_L) = \min[u_{A_1^k}(V_d), u_{A_2^k}(V_{oc}), u_{B^k}(I_L)], \quad k = 1, 2, \dots, n. \tag{7.35}$$
$$u_{B'}(I_L) = \max[u_{B^k}(I_L)] \quad k = 1, 2, \dots, n. \tag{7.36}$$

where $u_{B'}(I_L)$ denotes the aggregated output fuzzy set.

Table 7.7 Rule base of FL controller for linguistic variables.

Balancing current		V_d				
		NV	N	Z	P	PV
V_{oc}	VS	NV	NV	Z	Z	P
	S	NV	N	Z	P	PV
	M	NV	N	Z	P	PV
	L	NV	N	Z	P	PV
	VL	N	Z	Z	PV	PV

Step 4: Defuzzification of the output. This step converts the inference fuzzy output set to the crisp inductor current (I_L). The center of gravity method is used for defuzzification. It is given by:

$$I_L = \frac{\int u_{B'}(I_L) \cdot I_L dI_L}{\int u_{B'}(I_L) dI_L} \tag{7.37}$$

Step 5: Obtain the mode durations for each cell. With the inductor currents, the desired mode durations can be calculated by:

$$
\begin{pmatrix} d_1 \\ d_2 \\ d_3 \end{pmatrix}
=
\begin{pmatrix}
\frac{V_p}{R_{on}} - 2I_{L1} - I_{L2} & I_{L2} + I_{L3} & I_{L3} \\[2mm]
\frac{V_p}{R_{on}} - I_{L1} + I_{L3} & \frac{V_p}{R_{on}} + I_{L1} + I_{L3} & 2I_{L3} \\[2mm]
\frac{V_p}{R_{on}} + I_{L2} + 2I_{L3} & \frac{V_p}{R_{on}} + I_{L1} + I_{L2} + 2I_{L3} & \frac{V_p}{R_{on}} + I_{L1} + 2I_{L2} + 2I_{L3}
\end{pmatrix}^{-1}
$$

$$
\cdot
\begin{pmatrix}
\frac{V_{oc1}}{R_{on}} + I_{L1} + I_{L2} + I_{L3} + I_{L1}\frac{R_L}{R_{on}} \\[2mm]
\frac{V_{oc1} + V_{oc2}}{R_{on}} + I_{L1} + 2I_{L2} + 2I_{L3} + I_{L2}\frac{R_L}{R_{on}} \\[2mm]
\frac{V_{oc1} + V_{oc2} + V_{oc3}}{R_{on}} + I_{L1} + 2I_{L2} + 3I_{L3} + I_{L3}\frac{R_L}{R_{on}}
\end{pmatrix}
\tag{7.38}
$$

7.5.2.3 Adaptability of FL Controller

The FL controller has good adaptability in regulating the inductor current based on the feedback of V_{oc} and V_d. For example, if V_d is large, a high inductor current is desired; if V_d is small, a low inductor current is desired. The FL controller can adaptively change the inductor current for different situations.

Considering only V_d as the input with constant OCV at 3.3 V, the equivalent gain and the corresponding inductor current are shown in Figures 7.42 and 7.43, respectively. When V_d is positively or negatively large, the equivalent gain is small to keep the inductor current under the limit. As V_d becomes small, the equivalent gain increases. When V_d is about 25 mV, the equivalent gain reaches its peak value of 70, corresponding to a peak inductor current of 3 A.

When V_d continues decreasing to 15 mV, the equivalent gain starts to decrease sharply to less than 20 and the corresponding inductor current declines to near zero. This prevents the divergence of the already-balanced cells. For comparison, the gain of the PI controller is shown in Figure 7.42 as a dotted line, which results in the same peak inductor current.

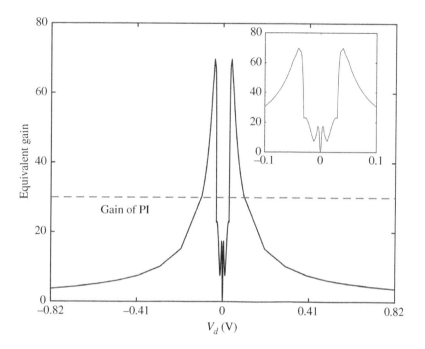

Figure 7.42 Equivalent gain for desired inductor current with input V_d at OCV of 3.3 V.

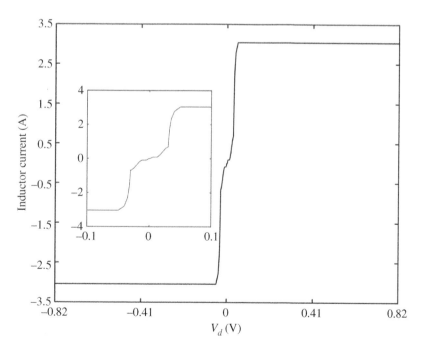

Figure 7.43 Inductor current with input V_d at OCV of 3.3 V.

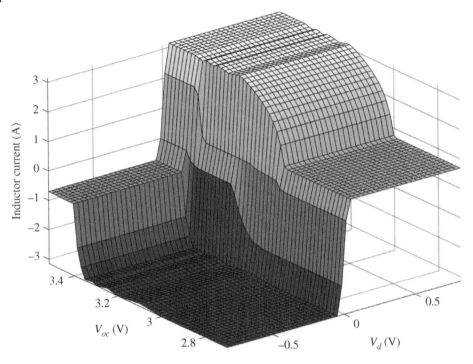

Figure 7.44 Inductor currents with two inputs of V_d and V_{oc}.

Considering both V_{oc} and V_d as inputs, the inductor currents are shown in Figure 7.44. When V_{oc} is higher than 3.5 V, representing a SOC higher than 90%, the inductor current is reduced and less balancing current is charged to this cell. When V_{oc} is lower than 3 V (about 20% of the SOC), the inductor current is also reduced, and less balancing current is discharged from this cell. This mechanism helps protect battery cells and potentially extend their service life.

7.5.2.4 Battery Balancing Criterion

To avoid the heavy computation burden of the SOC estimation, the OCV replaces the SOC as a balancing criterion in this study for the following two reasons: (i) the OCV directly reflects the SOC; and (ii) as a balancing criterion, the differences of the OCVs among the cells in the pack is more important than the absolute value of the OCVs. When the battery ECM as shown in Figure 7.2 is used, the OCV can be approximately calculated by

$$V_{oc} = V_b - V_{Diff} - V_R = V_b - I \cdot (R_{Diff} + R_i) \tag{7.39}$$

where V_{oc} represents the OCV. $R_{Diff} + R_i$ is obtained from the experiment and is taken as constant during the balancing process [9].

7.5.2.5 Experimental Results

The active balancing system based on the FL controller is evaluated and compared with the active balancing system based on the PI controller [23]. LFP battery cells from the A123 Company and NCA battery cells from Samsung are selected. After a screening

Table 7.8 Specifications of two types of battery cells.

Battery type	LFP	NCA
Manufacturing company	A123	Samsung
Nominal capacity (Ah)	2.3	2.5
Tested capacity (Ah)	2.13	2.5
Internal resistance (mΩ)	13	22

process, the four cells with almost the same internal resistances and capacities in each type are connected in series to make a battery pack. Initially, all four cells in each pack are fully charged and then one cell in each pack (e.g. cell four in this test) is discharged by 20% of the tested capacity to create an imbalance scenario. The specifications of these two types of battery cells are listed in Table 7.8. The frequency of the pulse width modulation (PWM) signal to control the MOSFETs in the MSI balancing circuit is 5 kHz. The inductor current and the mode duration decide the cell balancing currents based on Eq. (7.26). The highest inductor current and the peak battery balancing current are

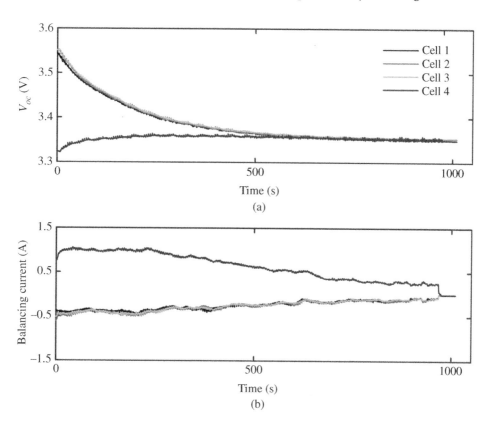

Figure 7.45 Experimental results for (a) OCV and (b) average balancing current of each cell in a LFP battery pack with FL controller.

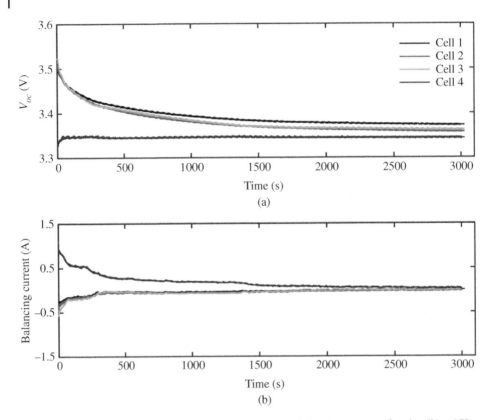

Figure 7.46 Experimental results for (a) OCV and (b) average balancing current of each cell in a LFP battery pack with PI controller.

set to 3 and 4 A, respectively. The balancing stops when the OCV difference is less than 10 mV or the balancing time is longer than 3000 seconds.

The LFP battery has flat voltage characteristics. Figures 7.45 and 7.46 show the OCVs and average balancing currents with the FL and the PI controllers for the LFP battery cells during the balancing process, respectively.

The balancing operation with the FL controller stops at around 1000 seconds when the maximum OCV difference reaches 10 mV. In contrast, the balancing operation with the PI controller stops at about 3000 seconds. This is because the FL controller maintains the higher balancing current for a longer time than the PI controller, which can be seen in Figures 7.45b and 7.46b, respectively. Furthermore, with the FL controller the three cells with the same initial OCVs keep converging as the gains among them are very small, as shown in Figure 7.45a. With the PI controller, until the balancing operation runs out of time, there are still significant OCV differences, as shown in Figure 7.46a due to the slow balancing speed as a result of the small average balancing currents in the balancing process.

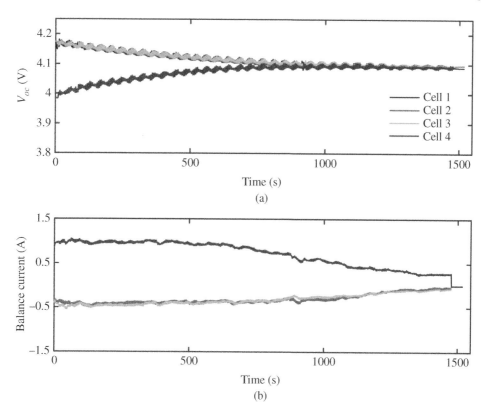

Figure 7.47 Experimental results for (a) OCV and (b) average balancing current of each cell in a NCA battery pack with FL controller.

In contrast with the LFP battery, the NCA battery has steep voltage characteristics. The OCVs and average balancing currents of the NCA batteries with the FL controller and the PI controller are shown in Figures 7.47 and 7.48, respectively. With the FL controller, the balancing ends at 1500 seconds when the maximum OCV difference reaches 10 mV. The balancing time for the NCA battery is slightly longer than that for the LFP battery, since the OCVs of the NCA battery drop slowly compared with those of the LFP battery. The recovered cell capacity is the highest in all cases. With the PI controller, the maximum OCV difference still cannot reach 10 mV within 3000 seconds. The OCVs drop slowly and the average balancing currents for the NCA battery are slightly higher than those for the LFP battery.

In the battery pack connected in series, the weakest cell decides the pack capacity. Therefore, the recovered pack capacity (energy) is the ratio of the increased capacity of the weakest cell to the tested capacity (energy). The balancing results for the A123 LFP battery pack and the Samsung NCA battery pack are shown in Tables 7.9 and 7.10, respectively. It can be seen that the FL controller has better balancing performance than

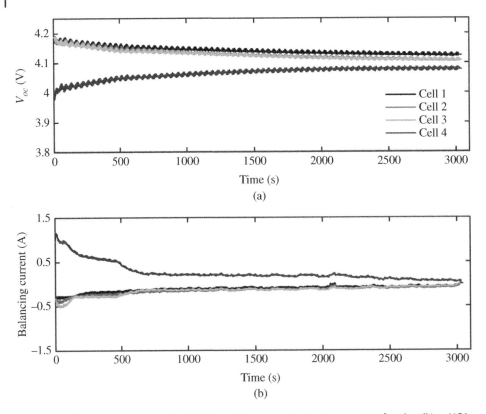

Figure 7.48 Experimental results for (a) OCV and (b) average balancing current of each cell in a NCA battery pack with PI controller.

Table 7.9 Balancing results for a LFP battery pack.

	PI controller		FL controller	
	Charge (Ah)	Energy (Wh)	Charge (Ah)	Energy (Wh)
Cell one	−0.0549	−0.1933	−0.0680	−0.2386
Cell two	−0.0680	−0.2387	−0.0717	−0.2510
Cell three	−0.0656	−0.2303	−0.0697	−0.2444
Cell four	+0.1299	+0.3675	+0.1758	+0.5327
Recovered capacity (energy) of cell four	6% (5.1%)		8.25% (7.4%)	

the PI controller in the LFP battery pack. With the PI controller, the final SOC difference is still 10% and 0.1299 Ah is charged to the weakest cell as the recovered pack capacity. With the FL controller, the final SOC difference narrows to 7.5% and 0.175 77 Ah is charged into the weakest cell as the recovered pack capacity.

Table 7.10 Balancing results for a NCA battery pack.

	PI controller		FL controller	
	Charge (Ah)	Energy (Wh)	Charge (Ah)	Energy (Wh)
Cell one	−0.0801	−0.3363	−0.1370	−0.575
Cell two	−0.1075	−0.4516	−0.1352	−0.5675
Cell three	−0.1067	−0.4478	−0.1325	−0.5562
Cell four	+0.2382	+0.8915	+0.2989	+1.1466
Recovered capacity (energy) of cell four	11.2% (10%)		14% (12.74%)	

7.6 Summary

In this chapter, various battery balancing methods are discussed, including battery sorting, passive balancing, and active balancing. Battery sorting is used in the initial state of making a consistent battery pack. The passive balancing and active balancing are used in the operation of the battery pack.

Two battery sorting methods are presented. One is to sort the battery cells into the group by directly comparing battery parameters of cells. The other is to sort the battery cells into the group using the SOM, where the input vector of the SOM is battery parameters of cells and the output vector of the SOM is the number of the group.

In passive balancing, the fixed shunt resistor, switched shunt resistor or switched transistor balancing circuits are presented to explain the working principle of passive balancing. It is only used in the battery pack charging process. In active balancing, balancing criterion, balancing control, and balancing circuits are discussed. The balancing circuits are divided into five groups based on the direction of energy flow between cells, energy storage tank and battery pack in the balancing process, including cell to cell, cell to pack, pack to cell, cell to energy storage tank to cell, and cell to pack to cell. It is used for the battery pack in both charging and discharging processes.

Two active balancing systems are used to demonstrate the capacity improvement of battery packs from the perspectives of selecting a balancing criterion and designing a balancing controller.

References

1 Hoque, M.M., Hannan, M.A., Mohamed, A., and Ayob, A. (2017). Battery charge equalization controller in electric vehicle applications. *Renewable and Sustainable Energy Reviews* 75: 1363–1385.

2 Li, X., Wang, T., Pei, L. et al. (2014). A comparative study of sorting methods for lithium-ion batteries. In: *IEEE Conference and Expo on Transportation Electrification Asia-Pacific*, 1–6. IEEE.

3 Kim, J., Shin, J., Chun, C., and Cho, B.H. (2012). Stable configuration of a li-ion series battery pack based on a screening process for improved voltage/SOC balancing. *IEEE Transactions on Power Electronics* 27 (1): 411–424.

4 He, F.X., Shen, W.X., Song, Q. et al. (2014). Clustering LiFePO$_4$ cells for battery pack based on neural network in EVs. In: *IEEE Conference and Expo on Transportation Electrification Asia-Pacific*, 1–5. IEEE.

5 Park, S.H., Park, K., Kim, H.S. et al. (2012). Single magnetic cell-to-cell charge equalisation converter with reduced number of transformer windings. *IEEE Transactions on Power Electronics* 27 (6): 2900–2911.

6 Lozano, J.G., Cadaval, E.R., Montero, M.M., and Martinez, M.A.G. (2014). Battery equalisation active methods. *Journal of Power Sources* 246: 934–949.

7 Cao, J., Schofield, N., and Emadi, A. (2008). Battery balancing methods: a comprehensive review. In: *IEEE Conference on vehicle power and propulsion*, 1–6. IEEE.

8 Qi, J. and Lu, D.C. (2014). Review of battery cell balancing techniques. In: *IEEE Australasian Universities Power Engineering Conference*, 1–6. IEEE.

9 Rahimi-Eichi, H., Ojha, U., Baronti, F., and Chow, M.Y. (2013). Battery management system: an overview of its application in the smart grid and electric vehicles. *IEEE Industrial Electronics Magazine* 7 (2): 4–16.

10 Vesanto, J. and Alhoniemi, E. (2000). Clustering of the self-organising map. *IEEE Transactions on Neural Networks* 11: 586–600.

11 Xiudong, C., Shen, W.X., Zhang, Y. et al. (2017). Novel active LiFePO$_4$ battery balancing method based on chargeable and dischargeable capacity. *Computers & Chemical Engineering* 97: 27–35.

12 Cassani, P.A. and Williamson, S.S. (2010). Design, testing, and validation of a simplified control scheme for a novel plug-in hybrid electric vehicle battery cell equalizer. *IEEE Transactions on Industrial Electronics* 57 (12): 3956–3962.

13 Cui, X.D., Shen, W.X., and Zheng, J. (2015). New on-line approach for lithium iron phosphate battery pack balancing based on state of charge. In: *IEEE International Conference on Electrical Machines and Systems (ICEMS)*, 762–767. IEEE.

14 Baughman, A.C. and Ferdowsi, M. (2008). Double-tiered switched-capacitor battery charge equalization technique. *IEEE Transactions on Industrial Electronics* 55 (6): 2277–2285.

15 Kim, M.Y., Kim, C.H., Kim, J.H., and Moon, G.W. (2014). A chain structure of switched capacitor for improved cell balancing speed of lithium-ion batteries. *IEEE Transactions on Industrial Electronics* 61 (6): 3989–3999.

16 Moo, C.S., Ching, H.Y., and Tsai, I.S. (2003). Charge equalization for series-connected batteries. *IEEE Transactions on Aerospace and Electronic Systems* 39 (2): 704–710.

17 Park, H.S., Kim, C.E., Moon, G.W. et al. (2007). Two-stage cell balancing scheme for hybrid electric vehicle lithium-ion battery strings. *IEEE Power Electronics Specialists Conference (PESC)* 273–279.

18 Einhorn, M., Roessler, W., and Fleig, J. (2011). Improved performance of serially connected Li-ion batteries with active cell balancing in electric vehicles. *IEEE Transactions on Vehicular Technology* 60 (6): 2448–2457.

19 Yarlagadda, S., Hartley, T.T., and Husain, I. (2013). A battery management system using an active charge equalization technique based on a DC/DC converter topology. *IEEE Transactions on Industry Applications* 49 (6): 2720–2729.

20 Imtiaz, A.M., Khan, F.H., and Kamath, H. (2011). A low-cost time shared cell balancing technique for future lithium-ion battery storage system featuring regenerative energy distribution. *Applied Power Electronics Conference and Exposition (APEC)* .

21 Karnjanapiboon, C., Jirasereeamornkul, K., and Monyakul, V. (2009). High efficiency battery management system for serially connected battery string. In: *IEEE International Symposium on Industrial Electronics (ISIE)*, 1504–1509. IEEE.

22 Cui, X.D., Shen, W.X., Zhang, Y.L., and Hu, C.G. (2017). A novel active online state of charge based balancing approach for lithium-ion battery packs during fast charging process in electric vehicles. *Energies* 10: 1766.

23 Cui, X.D., Shen, W.X., Zhang, Y.L., and Hu, C.G. (2017). A fast multi-switched inductor balancing system based on a fuzzy logic controller for lithium-ion battery packs in electric vehicles. *Energies* 10: 1034.

24 Simon, D. (2006). *Optimal State Estimation: Kalman, H Infinity, and Nonlinear Approaches*. Hoboken, NJ: Wiley.

25 Lim, C.S., Lee, K.J., Ku, N.J. et al. (2014). A modularized equalization method based on magnetizing energy for a series-connected lithium-ion battery string. *IEEE Transaction on Power Electronics* 29 (4): 1791–1799.

26 Wu, K.C. (2005). *Switch-Mode Power Converters: Design and Analysis*. Cambridge, MA: Academic Press.

27 Luo, M. and Zhang, K. (2015). Robustness of full implication algorithms based on interval-valued fuzzy inference. *International Journal of Approximate Reasoning* 62: 61–72.

28 Michels, K., Klawonn, F., Kruse, R., and Nürnberger, A. (2007). *Fuzzy Control: Fundamentals, Stability and Design of Fuzzy Controllers*, vol. 200. Berlin: Springer.

29 Ross, T.J. (2009). *Fuzzy Logic with Engineering Applications*. Hoboken, NJ: Wiley.

8

Battery Management Systems in Electric Vehicles

8.1 Background

A battery management system (BMS) is one of the core components in electric vehicles (EVs). It is used to monitor and manage a battery system (or pack) in EVs. The BMS ensures the safety and reliability of the battery system, improves its efficiency, prolongs its service life, and increases the driving range of EVs. Since a single cell has a limited capacity and voltage, a battery system is required to meet the needs of power and driving range of EVs. Generally, the battery system consists of a large number of battery cells connected in series or parallel or a combination of series and parallel. The BMS has a crucial role in managing such a battery system including the estimation of battery states and parameters and the prevention of battery overcharge, over-discharge, thermal runway, and battery inconsistences.

The rest of this chapter focuses on: (i) the composition and typical hardware of BMSs and their representative commercial products; and (ii) three key points of the next-generation BMSs: self-heating management, safety management of battery systems, and the application of cloud computation in BMSs.

8.2 Battery Management Systems

There are five main functions in terms of hardware implementation in BMSs for EVs [1]: (i) battery parameter acquisition; (ii) battery system balancing; (iii) battery information management; (iv) battery thermal management; and (v) battery charge control. Figure 8.1 shows a basic BMS hardware structure for EVs, which includes a digital core and several peripheral modules.

The digital core is a calculation unit which includes the estimation algorithms for the SOX (where X represents charge, energy, power and health, etc.) and control strategies. The peripheral modules include data acquisition and control circuits. The former is used to acquire the information of a battery system such as battery voltage, current, and temperature (analog signals) as well as the status of switches (digital signals). The latter is used for thermal and balancing management of a battery system and control of contactors inside a battery system or between the battery system and other parts of EVs. Figure 8.2 shows the basic BMS hardware circuits to implement these functions [2]. Battery charge has been discussed in Chapter 6 and its control can be realized by the information exchange between a BMS and a battery charger through the controller area

Advanced Battery Management Technologies for Electric Vehicles, First Edition. Rui Xiong and Weixiang Shen.
© 2019 John Wiley & Sons Ltd. Published 2019 by John Wiley & Sons Ltd.

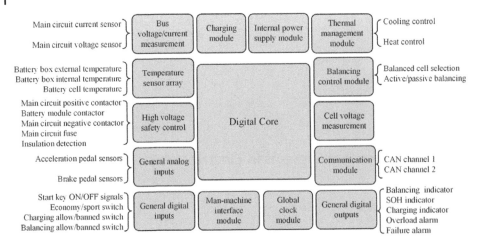

Figure 8.1 Basic BMS hardware structure for EVs.

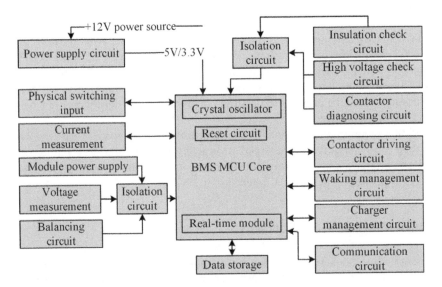

Figure 8.2 Basic BMS hardware circuits.

network (CAN). Therefore, we will only explain the first four functions of the BMS in details.

8.2.1 Battery Parameter Acquisition Module

To ensure the safety of the battery system, the voltage of each battery cell in the battery system needs to be sampled, which requires a large number of voltage acquisition channels. Two methods can be used to make this multi-channel acquisition circuit. One is to design a discrete circuit mainly consisting of a multiplexer, a differential circuit and an AD converter as shown in Figure 8.3. With a multiplexer, at any time only one analog signal among all multi-channel analog signals is connected to the AD converter through the differential circuit which is then converted into a digital signal. Currently, the voltage

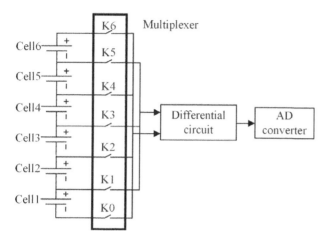

Figure 8.3 Discrete circuit of Multi-channel AD conversion for sampling battery voltage.

measurement accuracy ranges from 5 to 20 mV depending on the number of bits in an AD converter. The drawback of this method is that it is not possible to obtain all the sampled signals simultaneously for different battery cells in the battery system.

The other is to use the integrated circuit (IC) such as the LTC68xx series to design the voltage acquisition module. A schematic of LT6803-3 is shown in Figure 8.4 [3]. This IC can measure up to 12 series-connected battery cells. With serial interface daisy chains, multiple LTC6803-3 can be connected in series, allowing the voltage of all cells in the battery system to be measured. The microcontroller can communicate with this IC by a serial peripheral interface to obtain the voltages of each cell in a battery system. It takes 13 ms to measure the voltages of all 12 battery cells. The maximum measurement error is 0.25%. It supports different types of batteries and supercapacitors.

8.2.2   Battery System Balancing Module

The battery system balancing module implements balancing functions for the series-connected battery pack either by active balancing circuits or passive balancing circuits. For passive balancing circuits, Figure 8.5 shows an example of a switching shunt resistor balancing circuit [4]. By acquiring the voltage of the cells in the pack in real-time, the control circuit can determine the battery cell to be balanced by turning on the corresponding switch, allowing for the conversion of the excess energy into heat through the connected resistor. Recently, most commercial BMSs have adopted the passive balancing circuit for the following reasons [5]: (i) cells with similar electrochemical characteristics are selected to build a battery pack for EVs. When the consistent battery pack is in the cyclic operation, the inconsistency of the cells forms slowly and can be regularly eliminated by a passive balancing circuit during the charging process. (ii) Passive balancing only takes place during the charging process, energy loss has little effect on real operations of EVs. (iii) The passive balancing circuit is simple, reliable, and low cost. The LTC68xx series can be applied to design a passive balancing circuit. Take LTC6804-1 as an example [6]. For the 12th cell of the 12 series-connected cells in the battery pack, the passive balancing circuit parallels the 12th cell as shown in Figure 8.6.

The positive electrode and negative electrode of the 12th cell are connected to the pin of CELL12 and CELL11, respectively. Three wires on the right are connected to the pins

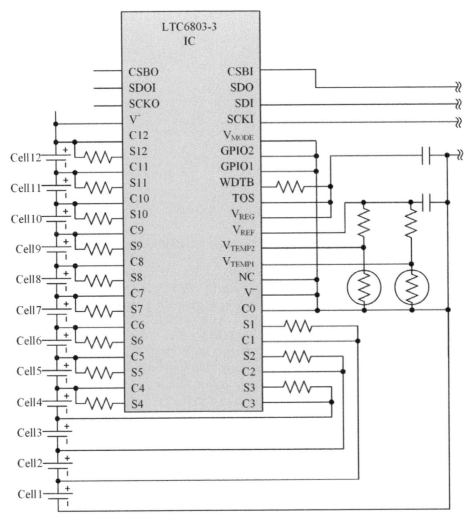

Figure 8.4 Voltage measurement using LT6803-3.

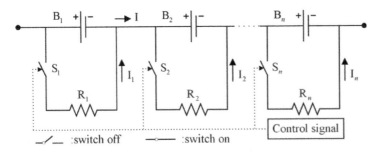

Figure 8.5 Battery passive balancing circuit with switching shunt resistor.

Figure 8.6 Passive balancing circuit.

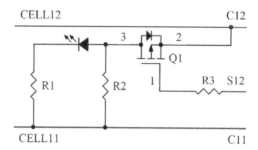

C12, C11, and S12 of the LTC6804-1 chip. The core of this circuit is R2, which is a 33 Ω dissipative resistor. The LTC6804-1 chip measures the voltage of the 12th cell through the pins of C12 and C11. When the 12th cell needs to be balanced, the chip controls the pin S12 to turn on the metal–oxide–semiconductor field-effect transistor (MOSFET). Then, the dissipative resistor will be connected to the 12th cell in parallel and the current flows through the resistor, allowing the excess energy to convert into heat.

The passive balancing circuit has been adopted in some commercial BMSs, such as the BMSs in EVs from BMW and Beijing Electric Vehicle Co. Ltd. (BJEV). Figure 8.7 shows an actual passive balancing circuit.

Active balancing circuits are mainly divided into three categories based on capacitances, inductors/transformers, and converters, respectively. The active balancing circuit based on transformers is widely used. Figure 8.8 shows an example of this balancing circuit [4]. It mainly consists of a primary winding, a primary switch, a number of secondary windings and branch switches. Suppose the battery B1 has more energy than the other batteries, this excess energy needs to be transferred to an entire battery pack. The two steps can complete this energy transfer. In step one, the primary switch K1 is switched on, the energy of B1 is stored in the magnetic field of the transformer. In step

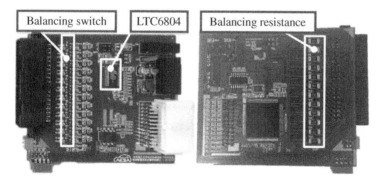

Figure 8.7 Actual passive balancing circuit.

Figure 8.8 Transformer-based active balancing circuit.

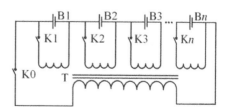

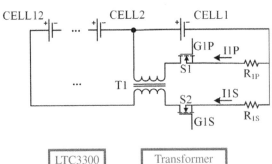

Figure 8.9 Active balancing circuit with the IC of LTC3300.

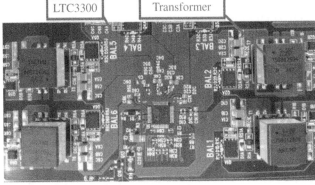

Figure 8.10 Actual active balancing circuit.

two, K0 is switched on and K1 is switched off, the energy stored in the transformer is transferred to the whole battery pack. These two steps can be repeated until all the cells in the battery pack are eventually balanced. For the battery with less energy, the above two steps can operate in reverse order to balance the cell.

This active balancing circuit can be designed by using the chip of LTC3300 [7], as shown in the Figure 8.9. In this balancing circuit, each battery cell is controlled by four corresponding pins, which are G1P, G1S, I1P, and I1S. The pins G1P and G1S control the MOSFETs S1 and S2 used as switches, respectively. The pins of I1P and I1S are used to measure the currents of the primary and secondary windings through the sampling resistances of R_{1P} and R_{1S}, respectively. Taking CELL1 as an example, when it has the excess energy, S1 is switched on and CELL1 charges the primary winding of the transformer T1. When the current of the primary winding increases to the preset maximum value, S1 is switched off and S2 is switched on, the energy stored in the transformer is transferred to the battery pack until the current of the secondary winding drops to zero. The above process can be repeated until energy transfer between the battery cell and pack is completed. Figure 8.10 shows the actual PCB for this active balancing circuit. Generally, the active balancing circuits are complicated, less reliable, and high cost.

8.2.3 Battery Information Management Module

The battery information management module has three main functions: (i) it stores key historic data of a battery system, such as SOX, the analysis of these key historic data can improve the estimation accuracy of the SOX; (ii) it displays key information which includes the current, voltage, temperature, SOX, and fault alarm; (iii) it communicates

Table 8.1 CAN communications in EVs.

Inner CAN	Control CAN	Instrument CAN	Debugger CAN
Cell voltage	Max cell voltage	Cell voltage	Error information
Cell temperature	Min cell voltage	Cell temperature	History information
SOC/SOH	Max temperature	Total voltage	Parameter information
Battery information	Min temperature	Total current	
Fault information	SOC/SOP/SOH	SOC	Version information
...	Fault information	Fault information	Inner CAN data
	Total voltage	...	Instrument CAN data
	Total current		...
	...		

between BMSs and other parts of EVs. The communication is mainly based on the CAN bus, including inner CAN communication between the battery monitoring circuit (BMC) and the battery control unit (BCU), the control CAN communication between the vehicle control unit (VCU) and the BMS, and the instrument CAN communication for a charger and a display and debugger CAN communication for software update and test. Different communication buses have different communication rates. A high rate is required for the inner CAN bus to ensure the frequency and synchronization of data acquisition, which is usually 250–500 kbps. Table 8.1 shows different CAN communications in EVs.

8.2.4 Thermal Management Module

The main functions of the thermal management module include: (i) measurement and monitoring of battery temperature; (ii) effective heat dissipation when the battery pack temperature is too high; (iii) rapid heating to suit operating temperatures of a battery system at low environmental temperature; and (iv) equalization of temperature distribution in a battery system. Because of the inconsistent heat dissipation, there exist temperature differences between cells in a battery system. This causes the inconsistent performances of the cells in the battery system which affects its capacity and shortens its life. In severe cases, the battery system may become thermally uncontrolled, leading to a serious accident.

Many cooling methods are proposed [8], such as air cooling, liquid cooling, and so on. Air cooling is a main method as it is easier to implement. Compared with air cooling, liquid cooling has a better cooling effect on the even temperature distribution of a battery pack. However, liquid cooling has high requirements for sealing the pack. If the liquid is conductive, it should be separated from the electric circuits. This increases the complexity of the cooling system and reduces the cooling effect to a certain extent.

Compared with battery cooling, battery heating has not received as much attention. Recently, the heating plate has been proposed to heat a battery pack in most EVs [9]. It has a simple structure, but it takes a long time to heat the battery pack to the preset temperature, causes uneven temperature distribution, and consumes high energy.

To improve the heating performance, the heating pipe has been proposed to reduce the heating time and allows the battery pack to be heated uniformly [10].

8.3   Typical Structure of BMSs

Depending on the number of cells in a battery system, BMSs can generally be divided into two categories: centralized and distributed [2]. The distributed BMS consists of the BMC and the BCU. In the distributed BMS, one BCU communicates with multiple BMCs by the inner CAN bus. The centralized BMSs integrate the BMC and the BCU on a single board, which connects to all the cells in a battery pack.

8.3.1   Centralized BMS

The centralized BMS implements the functions of acquisition, computing, and security monitoring, control management and communication on a single circuit board effectively, as shown in Figure 8.11. All the measurement signals are transmitted to a single circuit board. The advantage of the centralized BMS is that it can make use of the high speed of inter board communication to ensure the synchronized acquisition of parameters. The current sensor signals can also be collected directly without the CAN communication. However, the centralized BMS has the following disadvantages: (i) since all wiring harnesses are connected to a single circuit board, a large number of battery cells require a complicated design of connectors and wiring harnesses. Furthermore, it is difficult to protect the battery system when short circuit and overcurrent occur in different parts of the system. (ii) It is difficult to coordinate board size and high voltage security. Taking into account the high voltage safety issues, sufficient safety clearances must be ensured between the various channels, which leads to a large size of PCB. (iii) Since all the components are concentrated on a single board, this leads to poor scalability and maintainability. Furthermore, it is difficult to repair or replace components if they fail.

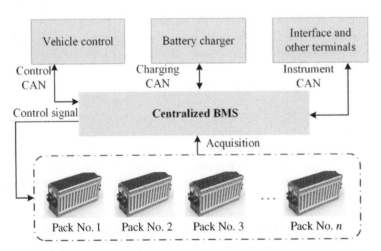

Figure 8.11 Architecture of centralized BMS.

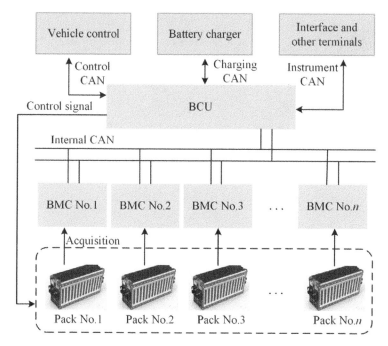

Figure 8.12 Architecture of distributed BMS.

8.3.2 Distributed BMS

The most commercial BMSs in EVs adopt the distributed structure, which consists of the BMC and the BCU, as shown in Figure 8.12.

The BMC is generally used for the acquisition of battery voltage, current, and temperature. It can also manage the safety and consistency of the battery pack. The BCU is used for fault detection, battery state estimation, switch management, and charge or discharge control. They are connected through an inner CAN. The distributed BMS greatly simplifies the structure with great applicability and scalability. However, it has the following disadvantages: (i) careful design is needed at the network level to ensure the integrity of communication and avoid communication failures. (ii) The distributed architecture means additional circuit and housing costs and each BMC needs to be individually numbered and programmed.

8.4 Representative Products

This section attempts to explain some of the commercial BMS products, such as E-Power, Klclear and Tesla, and some of the chips which can be used to design BMSs.

8.4.1 E-Power BMS

E-Power BMS has the representative products of EV05 series (http://www.hzepower .com/Products/ev05bcu.html). It adopts the distributed structure in this series, which includes the BCU and BMC shown in Figure 8.13.

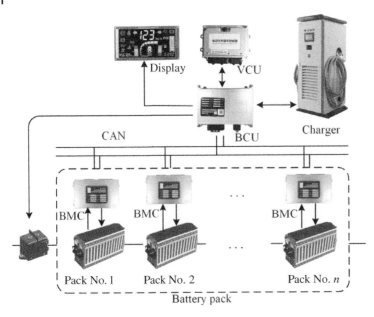

Figure 8.13 E-Power BMS architecture.

This BMS has the following features: (i) compatibility: it covers a full range of lithium-ion batteries (LiBs) including lithium manganese oxide (LMO), lithium nickel manganese cobalt oxide (NMC), lithium iron phosphate (LFP), and lithium titanate oxide (LTO) batteries for EV applications in commercial vehicles, passenger cars, and special vehicles. (ii) Standardization: the product software, hardware, and structural design are in accordance with the automotive standard. (iii) Scalability: it provides detection units that can support 30 BMCs at most with up to 300 batteries connected in series. (iv) Safety: the products are designed and tested in terms of system function self-test, high and low voltage resistance, protection grade, redundancy backup and anti-abuse. (v) Flexible configuration: the specially designed PC software allows the product to flexibly configure for different types of EV applications.

The E-Power BMSs have the functions of state of charge/state of health (SOC/SOH) estimation, battery malfunction analysis, balancing management (passive balancing with self-diagnosis function of balancing failure), thermal management (support heating and cooling management), safety protection (overcharge, over-discharge, over-current, over temperature, insulation leakage and smoke), and so on. Table 8.2 lists the technical specifications of this product.

8.4.2 Klclear BMS

Klclear BMSs provide active balancing function with balancing current up to 2A (http://www.klclear.com/index.php?a=shows&catid=6&id=17). They can be applied to LMO, NMC, LFP, and LTO batteries. There are no limitations to the minimum numbers of battery cells connected in series. One of their representative BMS products is the BMS05 series and adopts the centralized structure, as shown in Figure 8.14. It has the following three features: (i) it integrates the functions of cell voltage acquisition, bus insulation detection and relay fault detection. (ii) it has up to 48 channels for voltage acquisition,

Table 8.2 Technical specifications of E-power BMS.

Parameter	Value
BMS operating temperature range	$-40{\sim}85\,°C$
Single cell voltage detection range	$0{\sim}5\,V$
Single cell voltage sampling accuracy	$\leq\pm10\,mV\,(-25{\sim}55\,°C)$
Single cell voltage sampling frequency	$<20\,ms$
Total voltage measurement range	$0{\sim}900\,V$
Temperature measurement range	$-40{\sim}125\,°C$
Temperature detection accuracy	$\leq\pm1\,°C\,(NTC,@0{\sim}65)$
Current detection accuracy	$<1\%$ (Full scale range)
SOC estimation accuracy	$<8\%$
Passive balancing current	$100\,mA$

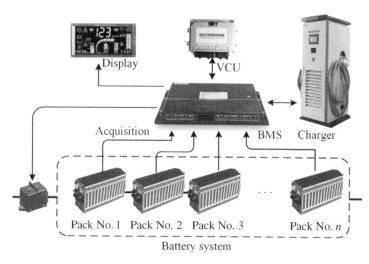

Figure 8.14 Klclear BMS architecture.

8 channels for temperature acquisition, and 3 channels for CAN communication. The measurement accuracy for a single cell voltage is less than 10 mV and that for a current is less than 0.5%. (iii) it is embedded with the general packet radio service (GPRS) function for remote monitoring. (iv) it supports the CAN network wake-up, online CAN upgrade, and UDS protocol upgrade. The UDS refers to the unified diagnostic services (UDS) protocol upgrade which is defined in ISO14229 [11].

8.4.3 Tesla BMS

Tesla's battery pack consists of a huge number of 18 650 Li-ion batteries, which makes the BMS extremely complicated. Consider Tesla Model S as an example, each car has more than 5000 battery cells, which consists of up to 84 packs connected in series (https://www.zhihu.com/question/22183103).

Table 8.3 Typical battery management ICs.

Company	Part number	Description
Linear	LTC6811	0.025% measurement accuracy; built-in isolated communication (isoSPI); 16-bit ADC; 12 voltage acquisition channels; 290 µs sampling period
Maxim	MAX14921	Up to 16 voltage acquisition channels; less than ±0.5 mV error; support passive balancing; open-wire and short-circuit fault detection; under/overvoltage warning
TI	BQ76PL455A	Up to 16 voltage acquisition channels; support passive balancing and active balancing with EMB1428Q/EMB1499Q; integrated 14-bit ADC; over-temperature, overvoltage, and undervoltage warning

The Tesla BMS has a distributed structure. Each battery pack is equipped with a BMC. The BCU realizes high voltage and insulation detection, high voltage interlock, contactor control, external communication, and other functions. The BMC is responsible for detecting the voltage and temperature of the batteries and reporting it to the BCU by the universal asynchronous receiver/transmitter (UART) communication. The Tesla BMS adopts a dual microcontroller unit (MCU) design. The secondary MCU can detect the working status of the master MCU. Once the master MCU fails, the secondary MCU can gain control authority to avoid any danger. The BMS uses a passive balancing circuit with a balancing current of about 100 mA.

8.4.4 ICs for BMS Design

The ICs for the BMS design are highly integrated devices to meet the functional requirements of most BMSs [12–14]. They simplify the design of BMSs and improve their miniaturization and reliability. Table 8.3 lists some of the ICs and their performances.

8.5 Key Points of BMSs in Future Generation

The current BMSs have achieved basic functions such as the SOC estimation, SOH estimation, balancing management, and thermal management. Due to the factors such as a large number of battery cells in a battery system, nonlinear characteristics and inconsistent aging trajectories of battery cells and complicated EV operating conditions, it is necessary for the development of next-generation BMSs to meet future demand. The next-generation BMS should be equipped with the following new features:

1) A rapid heating function to meet EV applications for all climate environments.
2) A diagnosis function to identify battery faults and battery sensor faults for the improved safety performances of a battery system.
3) Cloud computing maximizes the use of different data resources including the operational data of a battery system. Its applications have the advantages of high computing power, data storage security, and data sharing. Combining cloud computing and on-board control can greatly improve the performance of BMSs.

8.5.1 Self-Heating Management

To meet EV applications for all weather conditions, a battery pack in a low temperature environment needs to be preheated. There are external and internal heating methods [15]. External heating methods use air, liquid, or phase change material to preheat a battery pack. Internal heating methods use AC currents, internal resistive heating, convective heating and mutual pulse heating to preheat a battery pack. They are generally time and energy-consuming with complicated heating circuits and less applicable to a large capacity battery [16].

Recently, self-heating has been proposed to embed the nickel foil inside a LiB, enabling a rapid and efficient heating process, where the internal temperature is controlled by an internal temperature sensing circuit [17]. The heating schematic is shown in Figure 8.15. Different from conventional LiBs, this new LiB is constructed to include an activation terminal to control self-heating. In addition to three basic components (node, cathode, and electrolyte), a fourth component is added to the battery: a nickel (Ni) foil 50 μm thick, which has two tabs at each end. One tab is electrically jointed to the negative electrode. The other tab, as the activation terminal, is used to activate battery internal heating at low temperature. A switch connects the activation terminal with the negative electrode. When the switch opens, electrons will flow through the Ni foil, generating substantial ohmic heat, which rapidly warms up the core of the battery. When the switch closes, the activation process is completed. The time required for the battery to increase from −20 to 0 °C is 20 seconds and the power consumption is 3.8% capacity. This self-heating method is more practical and promising in EV applications.

Figure 8.15 Schematic of nickel foil-based self-heating.

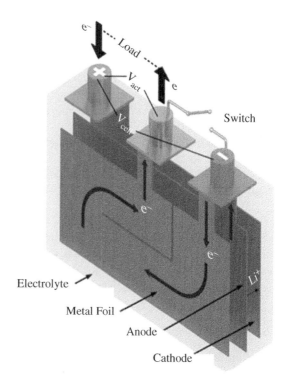

8.5.2 Safety Management

Battery faults induce safety issues in EVs. Among all the battery faults, short circuit faults can cause serious or catastrophic accidents [18]. The short circuit faults produce a large amount of heat in a very short period of time which spread quickly to the surroundings of a battery system. This can easily lead to fires or even disastrous explosions. Short circuit faults can be classified as internal short circuit (ISC) faults and external short circuit (ESC) faults.

The ISC faults in a battery cell may be caused by physical abuse (e.g. drop and crush) or operational abuse (e.g. overcharge by a poorly designed charger) or manufacturing defects (e.g. metallic contaminants, electrode misalignment) [19]. The ISC triggers a local hot spot which can propagate throughout an entire battery pack, leading to thermal runaway and fire [20]. The relationship between the ISC and the parameters of a battery ISC equivalent circuit model (ECM) needs to be established. If the voltage and temperature in the period of the ISC incubation have a strong relationship with the ISC model parameters, the ISC can be inferred and detected by identifying the parameters of the ISC model [21]. The model-based parameter identification algorithm (such as recursive least squares, RLS, with forgetting factor) can be employed to transform the voltage and temperature responses into physical parameters, whose values are used to interpret the ISC status. Besides, the self-discharge phenomenon caused by the ISC is considered as another method to achieve ISC detection [22].

The ESC faults in a battery cell are caused by accidently connecting a very low resistance between the positive and negative electrodes of the battery cell mainly due to falling off or dislocation of surrounding conductive components. The ESC fault easily induces one cell exothermic side reaction which affects other surrounding cells and triggers a chain of exothermic reactions [23]. Generally, when the ESC occurs, the BMS can capture only three quantities: current, voltage, and temperature, where the current reaches a very high level and the voltage tends to be zero. Thus, the ESC can be simply diagnosed by setting current and voltage thresholds. Although the temperature rise has a time lag, a recent study shows that the temperature during the ESC fault can increase very rapidly which requires fast and accurate detection of the ESC fault [24]. The battery ESC ECM is introduced to improve the diagnostic precision [25]. Once the current or voltage of the cell is abnormal, the predefined ESC model is activated. The root mean squared error (RMSE) of battery voltage between the model prediction and experimental data is then evaluated for fault prognosis. If the RMSE values are larger than the preset threshold, the fault is detected. However, improper setting of the RMSE threshold can easily lead to erroneous diagnosis. Accordingly, a multilevel RMSE threshold is set for different ESC phases.

Recent BMSs only measure battery parameters such as current, voltage, and temperature; these battery parameters make early detection of short circuit faults difficult. The development of early detection algorithms and their fast and accurate hardware implementation is required for use in future BMSs in EVs.

8.5.3 Cloud Computing

With the development and application of information technology, cloud computing is gradually being applied in BMSs for EVs. A cloud computing-based system that is

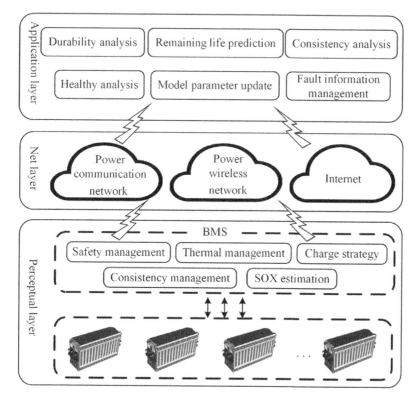

Figure 8.16 Architecture of cloud computing for BMSs in EVs.

combined with an on-board controller can greatly improve the performances of BMSs. Its advantages are reflected in the following three aspects:

1) *Intelligent and powerful algorithms.* Battery SOX algorithms and control strategies are computationally intensive. They are difficult to be implemented in an on-board controller in EVs. With the cloud computing platform, the computing performance of the BMS can be greatly improved.
2) *High calculation accuracy.* High computing speed, parallel computing and large storage space can implement multi-algorithms, which improve the calculation accuracy.
3) *High security.* Cloud computing can realize multiple models and multiple algorithms running at the same time. The system has high redundancy, and it is easy to extend functions such as fault diagnosis, and integrates big data analysis to achieve integrated analysis of security.

According to the structure of the Internet of Vehicles [26], we can propose the architecture of cloud computing for BMSs in EVs, as shown in Figure 8.16. It mainly includes three parts: the perceptual layer; the net layer; and the application layer.

The perceptual layer collects cell voltage, current, and temperature to calculate the SOX for safety management, thermal management, consistency management, and charge management. It is expected that there will be only one chip embedded with all the SOX estimation algorithms. This chip has the following features:

1) Sample and calibrate the measured data.
2) Estimate battery states such as the SOC, SOH, state of power (SOP), and state of energy (SOE).
3) Diagnose the working status of the cell and upload the safety status in time.
4) Maintain the consistency of the cells to prolong the life of a battery system.
5) Realize the function of remote transmit-receive.
6) Isolate the battery's physical signals electrically.
7) Consume low power.

The net layer is responsible for communications and data transfer. It includes the power communication network (PCN), the power wireless network, and the Internet. The PCN is an important infrastructure of the power system [27]. When the EV is under charging, the charger can communicate with the PCN to send some battery information [28]. The power wireless network is the radio access technology, e.g. cellular system, Wi-Fi, or satellite [29]. It will work with a remote data monitoring module installed in EVs and transfer all battery data in EVs to data centers.

The application layer is the core of cloud computation. It can achieve six main functions: battery durable analysis, battery remaining life estimation (RUL), battery consistency analysis, battery health analysis, battery model parameter calibration, and faults information management. Consider battery RUL prediction and consistency analysis as examples to illustrate the application of cloud computing in BMSs for EVs.

For RUL prediction, cloud computing can combine online experimental data and a large number of battery history degradation data (HDD) with the deep learning method. This cloud computing-based RUL prediction requires high computation capability (such as a large-scale neural network with distributed computing) supported by high storage, which is hard to be implemented in current BMSs. Figure 8.17 shows the flowchart of RUL prediction based on deep learning [30]. In this method, the RUL prediction is considered as a time series prediction problem. A LSTM-RNN is employed to learn the aging trajectory from the battery HDD and a multi-step-ahead prediction (MSAP) is performed by giving the current capacity until the end of life. Monte Carlo simulation is used to generate the prediction uncertainties to make sure that the LSTM-RNN provides an accurate and reliable RUL prediction.

For the inconsistency of cells in a battery pack, the current BMS cannot store long-time history data of the cells because of its low capability of computation and storage. The next-generation BMS with cloud computing can combine a large amount of calibration data stored in the cloud and the dynamically assigned storage memory to identify inconsistency of the cells in the pack through equivalent circuit of the cell difference model [31]. This can be embedded in the application layer of cloud computing system.

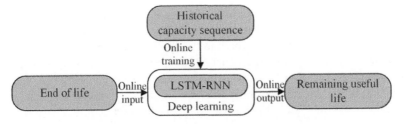

Figure 8.17 Flow chart of RUL prediction based on deep learning.

8.6 Summary

This chapter has introduced the main functions of BMSs including battery parameter acquisition, battery system balancing, battery information management and thermal management, and the two typical structures of BMSs (centralized BMS and distributed BMS).

The representative BMS products from various companies are discussed. To meet the future demand of BMSs in EVs, the self-heating approach, the safety management strategy and cloud computing are recommended in the next generation of BMSs.

References

1 Lu, L., Han, X., Li, J. et al. (2013). A review on the key issues for lithium-ion battery management in electric vehicles. *Journal of Power Sources* 226: 272–288.

2 Lu, K. (2018). Hardware architecture development of battery management system in electric vehicles. *Electronic Engineering and Product World* 25 (05): 28–31.

3 Linear Technology Corporation. LTC6803-3 datasheet. http://www.analog.com/media/en/technical-documentation/data-sheets/680313fa.pdf.

4 Daowd, M., Omar, N., Bossche, P.V.D., and Mierlo, J.V. (2011). Passive and active battery balancing comparison based on MATLAB simulation. In: *IEEE International Conference on Vehicle Power and Propulsion*, 1–7. IEEE.

5 Xiong, R. and Duan, Y. (2018). Development & verification of the balance strategy for batteries in electric vehicles. *Journal of Beijing Institute of Technology* 27 (01): 22–28.

6 Linear Technology Corporation. LTC6804-1 datasheet. https://pdf1.alldatasheet.com/datasheet-pdf/view/495163/LINER/LTC6804-1.html.

7 Linear Technology Corporation. LTC3300 datasheet. http://www.analog.com/media/en/technical-documentation/data-sheets/33001fb.pdf.

8 Madani, S.S., Swierczynski, M.J., and Kær, S.K. (2017). A review of thermal management and safety for lithium ion batteries. In: *IEEE International Conference on Ecological Vehicles and Renewable Energies*, 1–20. IEEE.

9 Jarrett, A. and Kim, I.Y. (2014). Influence of operating conditions on the optimum design of electric vehicle battery cooling plates. *Journal of Power Sources* 245 (1): 644–655.

10 Wang, Q., Jiang, B., Xue, Q.F. et al. (2015). Experimental investigation on EV battery cooling and heating by heat pipes. *Applied Thermal Engineering* 88: 54–60.

11 Wajape, M. and Elamana, N.B. (2014). Study of ISO 14229-1 and ISO 15765-3 and implementation in EMS ECU for EEPROM for UDS application. In: *IEEE International Conference on Vehicular Electronics and Safety*, 168–173. IEEE.

12 Maxim Integrated. MAX14921 datasheet. https://datasheets.maximintegrated.com/en/ds/MAX14920-MAX14921.pdf

13 Texas Instruments. BQ76PL455A datasheet. https://pdf1.alldatasheet.com/datasheet-pdf/view/807921/TI1/BQ76PL455A-Q1.html.

14 Linear Technology Corporation. LTC6811-1 manual. http://www.analog.com/media/en/technical-documentation/data-sheets/68111fb.pdf.

15 Zhang, G., Ge, S., Xu, T. et al. (2016). Rapid self-heating and internal temperature sensing of lithium-ion batteries at low temperatures. *Electrochimica Acta* 218: 149–155.

16 Stuart, T.A. and Hande, A. (2004). HEV battery heating using AC currents. *Journal of Power Sources* 129 (2): 368–378.

17 Wang, C.Y., Zhang, G., Ge, S. et al. (2016). Lithium-ion battery structure that self-heats at low temperatures. *Nature* 529 (7587): 515.

18 Feng, X., Ouyang, M., Liu, X. et al. (2017). Thermal runaway mechanism of lithium ion battery for electric vehicles: a review. *Energy Storage Materials* 10: 246–267.

19 Cai, W., Wang, H., Maleki, H. et al. (2011). Experimental simulation of internal short circuit in Li-ion and Li-ion-polymer cells. *Journal of Power Sources* 196 (18): 7779–7783.

20 Kim, G.H., Pesaran, A., and Spotnitz, R. (2007). A three-dimensional thermal abuse model for lithium-ion cells. *Journal of Power Sources* 170 (2): 476–489.

21 Feng, X., Weng, C., Ouyang, M., and Sun, J. (2016). Online internal short circuit detection for a large format lithium ion battery. *Applied Energy* 161: 168–180.

22 Seo, M., Goh, T., Park, M. et al. (2017). Detection of internal short circuit in lithium ion battery using model-based switching model method. *Energies* 10 (1): 76.

23 Orendorff, C.J., Roth, E.P., and Nagasubramanian, G. (2011). Experimental triggers for internal short circuits in lithium-ion cells. *Journal of Power Sources* 196 (15): 6554–6558.

24 Chen, Z., Xiong, R., Tian, J. et al. (2016). Model-based fault diagnosis approach on external short circuit of lithium-ion battery used in electric vehicles. *Applied Energy* 184: 365–374.

25 Yang, R., Xiong, R., He, H., and Chen, Z. (2018). A fractional-order model-based battery external short circuit fault diagnosis approach for all-climate electric vehicles application. *Journal of Cleaner Production* 187: 950–959.

26 Papadimitratos, P., Arnaud, D. L. F., Evenssen, K., Brignolo, R., & Cosenza, S. (2009). Vehicular communication systems: enabling technologies, applications, and future outlook on intelligent transportation. Communications Magazine IEEE, 47(11), 84–95.

27 Xie, Y., Wang, Y., Wei, L.I. et al. (2017). Indicator system for operation quality of electric power communication network. *Electric Power* 50 (10): 22–27.

28 Ren, K.M., Ji-Zhou, L.I., Liu, L.Y., and Song, W.Y. (2015). Development status and tendency of internet of vehicles communication technology. *Communications Technology* 48 (5): 507–513.

29 Motlagh, N.H., Taleb, T., and Arouk, O. (2016). Low-altitude unmanned aerial vehicles-based internet of things services: comprehensive survey and future perspectives. *IEEE Internet of Things Journal* 3 (6): 899–922.

30 Zhang, Y., Xiong, R., He, H., and Pecht, M. (2018). Long short-term memory recurrent neural network for remaining useful life prediction of lithium-ion batteries. *IEEE Transactions on Vehicular Technology* https://doi.org/10.1109/TVT. 201 8.2805189.

31 Ouyang, M., Zhang, M., Feng, X. et al. (2015). Internal short circuit detection for battery pack using equivalent parameter and consistency method. *Journal of Power Sources* 294: 272–283.

Index

Advanced Battery Management Technologies for Electric Vehicles, First Edition. Rui Xiong and Weixiang Shen.
© 2019 John Wiley & Sons Ltd. Published 2019 by John Wiley & Sons Ltd.

Printed and bound by CPI Group (UK) Ltd, Croydon, CR0 4YY

09/02/2023

03190625-0002